INSECTS DID IT FIRST

Greg Paulson, E. Paul Catts, and Roger D. Akre indicate that insects are indeed 1st.

INSECTS DID IT 1ST.

by
Roger D. Akre,
Gregory S. Paulson
and   E. Paul Catts

Illustrations by
E. Paul Catts

Ye Galleon Press
Fairfield, Washington

Library of Congress Cataloging-in-Publication Data

Akre, Roger D., 1937
    Insects did it first / by Roger D. Akre, Gregory S. Paulson, and
E. Paul Catts ; illustrations by E. Paul Catts.
        p. cm
    Includes bibliographical references (p.   ) and index.
    ISBN 0-87770-517-8
    1. Insects, I. Paulson, Gregory S. II. Catts, E. Paul (Elmer
Paul) III. Title.
QL463.A38 1992
595.7--dc20   92-38722
                                                                          CIP

Dedication

RDA -This book is an attempt to interest students of natural history in the fascinating and incredible lifestyles of the most abundant creatures on earth, the insects. I dedicate this book to these students and to the two women who influence my life daily.

GSP - I would like to thank my parents, Mr. and Mrs. Neil A. Paulson, Sr., for their support and encouragement through the years. I'd also like to acknowledge the great influence that Dr. Sally L. Paulson and Mr. Neil A. Paulson, Jr., my sister and brother, had on my life. Where would I be today without the Bee Club?

EPC - I dedicate my effort in this book to Dale F. Bray and Walter A. Connell, former faculty at the University of Delaware, who first introduced me to the astounding and fascinating science of entomology.

About the Authors

Roger D. Akre - Dr. Akre is a professor and entomologist at Washington State University who conducts research on several urban insect/arthropod pests including yellowjackets, carpenter ants, and venomous spiders. He has published many extension bulletins, and has a large clientele who frequently seek identifications of arthropods and advice about insect problems. He teaches insect behavior, insect photography, and the use of insects in teaching.

Gregory S. Paulson - Dr. Paulson's career in entomology has been devoted to the applied side of the science. He is especially interested in developing alternatives to pesticides for insect control. He has served as a Peace Corps volunteer in Western Samoa in a WHO filariasis research program and studied plant pathology in Hawaii. Most recently he has studied ant population structure in pear orchards. Presently, he is a biology instructor at Washington State University.

E. Paul Catts - Dr. Catts is a professor and entomologist at Washington State University specializing in medical, veterinary, and forensic entomology. His research deals with bot flies, biting flies, and flies that frequent decomposing carrion. Besides teaching medical entomology and insect morphology, he teaches a popular course entitled "Insects and People".

Table of Contents

Insects Did It First

Preface

The original ideas on "insects did it first" started with Dr. Roger Akre's first class in general entomology at Washington State University in 1964. Each time some advanced human technology, such as radar or sonar, became the topic of the day Roger realized that "people aren't really all that original - insects did it first." Soon he started writing down all the ideas that occurred to him on this topic, even in the middle of lectures, be it general entomology, insect behavior, or insect morphology.

Dr. Akre also started to make color slides of insects with the idea that perhaps he could someday make "first's" a special topic in his general entomology class. He was aided in his endeavors by several colleagues with artistic talents, Paul Catts and Robert Harwood, both entomologists at Washington State University, and by Dr. William B. Garnett, from the University of Cincinnati. All produced cartoons of the topics for slide lectures. The topics grew and interest was expressed by a great number of people, especially those who give insect talks to young scientists from the age of 3 to 30. They wanted copies of the slides and references to the material to create their own talks.

The thought of publishing "First" as a book occurred to Roger in 1986. By this time, students, colleagues, and friends were well acquainted with his interests in this area, and one, Greg Paulson, even located a book in 1987 that was similar in topic to the contents of this book. Lucy Berman and Roy Combs published "Wonderbaarlijke Nature" in Europe in 1971. In 1972 an English translation of this book, entitled "Nature Thought of It First", was printed by Grosset and Dunlop. Their book covers all animals, while ours concentrates solely on insects, treating them in much greater detail. Still another book, with nearly identical ideas and even a close title, "Nature Invented It First", was authored by R. E. Hutchins (1980).

"Insects Did It First" is organized into 81 chapters, each concerns a different achievement in which insects have precedence over other animals, including humans. Each chapter includes references to books and scientific papers in which the achievement or behavior is described. In most instances, several different insects are used as examples to illustrate the variety and extent of the "first". Each chapter is illustrated with a cartoon and, in many cases, a photograph. When appropriate, the Order and Family of the insect are included in the text. Family names can be easily recognized because they always end in -idae. A complete list of the insect orders can be found on page 140. Several topics in this book deal with sexual or other topics that some people might find sensitive, especially in regard to younger grade school children. With this in mind we placed those chapters at the end of the book so they can easily be excluded if this is deemed appropriate.

Introduction

Insects are the most numerous and diverse group of animals on the earth today. We don't know how many different kinds (**species**) of insects exist, but there are millions! There are probably more than one million species of beetles alone. In a typical backyard there can be more than a thousand species of insects with a population that totals several hundred thousand. Insects occur naturally in almost every terrestrial and freshwater habitat and, with the help of humans, insects have invaded Antarctica, sailed the high seas in tall ships, dove to the ocean's depths in submarines, and experienced the weightlessness of space.

Scientists, called **taxonomists**, study the relationship, or grouping, of species into categories (**taxa**, singular **taxon**). Species within each taxon share a set of characteristics that define each group and yet each has its own unique characteristics too. The first major taxon is **Kingdom**. Most biologist use the **five kingdom system** (Animalia, Plantae, Fungi, Protista, and Monera). Each kingdom is subdivided into **phyla** (singular **phylum**) or **divisions** (a term used for plants, fungi, and algae). Insects (Kingdom Animalia) are a major part of the **Phylum Arthropoda** (arthro=jointed or segmented, poda=foot or appendage) and also of the next smaller subgroup **Class Hexapoda** (hexa=six, "six-legged") (**Insecta**). In addition to hexapods, other classes of arthropods include the extinct trilobites (fossils only), horseshoe crabs, crustaceans (shrimp, lobsters, crabs), arachnids (scorpions, spiders, mites, ticks), millipedes, centipedes, and a few other less common animals.

Classes are further divided into **orders**. Class Hexapoda includes 28 orders of insects. Most people can easily distinguish orders of insects. For example, all beetles are in the Order **Coleoptera**, and all butterflies and moths are members of the Order **Lepidoptera** (a complete list of insect orders is located on page 140). It becomes a little more difficult to identify the subdivisions of orders, called **families**, without some training, but most people would be surprised at how many families they can recognize. For example, even though ladybird beetles ("ladybugs"), weevils, and click beetles share characteristics that identify them as Coleoptera, each of these groups is a distinct family, and most people can recognize that there are differences among them. Several excellent field guides, available at most bookstores, can provide a starting point for learning to identify insect orders and families (Borror & White 1970, Milne & Milne 1980). Within the families are the **genera** (singular **genus**) which consists of groupings of related species.

Fossils indicate that insects have inhabited the earth for at least 350-400 million years, about 200 times longer than humans! During this time they have evolved into an incredible array of species. Insects range in size from 1/100th to 13 inches in length, and about 1/50th to 12 inches in wingspan. They feed on a seemingly endless variety of food, each species has its own dietary needs, and they feed in many different ways (Borror et al. 1989). Some insects cause enormous problems by detrimentally affecting our health and causing losses to crops and livestock, stored products, and structures. However a greater number of insects are benefactors to humans. Insects are indespensible pollinators; produce honey, silk, and other valuable products; perform essential services such as scavenging and controlling harmful insects; and have an prominent role in the complex web of life. It has been said that without insects the world, as we know it, would cease to exist in less than a year.

Despite the variety of forms and life styles of insects they all possess several common features as adults that distinguish them from other groups of animals. Insects have a segmented body grouped into three main body regions - **head, thorax,** and **abdomen.** The head, containing the brain, has one pair of **antennae,** one pair of **mandibles** ("jaws"), and two pairs of **maxillae** (mouth parts). The second pair of maxillae are fused into a "lower lip" called the **labium.** The thorax is composed of 3 segments (the **pro-, meso-,** and **metathorax**) each of which bears a pair of jointed legs. In addition, the meso- and metathorax may each have a pair of wings. Insects and all other arthropods have a jointed **exoskeleton** (skeleton on the outside of the body) made of

2

chitin, a very durable, colorless, flexible material. Hard insects have an exoskeleton which contains a lot of **sclerotin**, a very hard protein/chitin compound.

There is incredible and fascinating diversity in almost every aspect of the biology of insects. It is therefore beyond the scope of this book to discuss general topics, but there are many good textbooks dedicated solely to **entomology**, the study of insects (Borror et al. 1989, Chapman 1982, Elzinga 1987, & Gillott 1970). The purpose of this book is to examine only certain aspects of insect biology and behavior, specifically, those capabilities which insects developed long before any other animals, including us.

1.   Architects

People have always been builders, from the construction of crude shelters to the incredibly tall skyscrapers of this modern age. On a much smaller scale however, insects are also builders, and their constructions are equally incredible. For example, immature caddisflies (Trichoptera) make cases around their bodies for protection (see Chapter 46). The cases not only serve to disguise the insects as another extraneous piece of the environment, but the case also gives its maker some degree of physical protection. This is true when the case is made of wood or plant debris, but it is especially strong when composed of sand grains or small rocks. Cases are also used by other Insects including some butterflies and moths (Lepidoptera) larvae (Borror et al 1989).

Ants (Hymenoptera: Formicidae) often make remarkably complicated nests, and common among these are mounds of the thatching ants, genus *Formica*. These ants are important **biological control agents** of insect pests in temperate forests, and in most forests their mounds are everywhere. The mounds are constructed of grass stems and other plants cut into short pieces. The mound is constantly being reshaped to catch the maximum amount of sunlight, to shed water from rain, or to increase air circulation. Many other ants have equally ingenious nests (Holldobler & Wilson 1990).

Like ants, all termites (Isoptera) are eusocial (see Chapter 27), however most are limited in distribution to the tropics and sub-tropics. Members of the Macrotermitidae, a Family of termites found in Africa, make huge, complex nests. These are fungus growers (see Chapter 17), and the mounds of one species, *Macrotermes bellicosus* (Smeathman), can have a diameter of about 32 yards (30 meters) (Wilson 1971). Another member of this group, *Macrotermes natalensis* (Haviland), makes nests up to 16 feet (5 meters) in diameter at the base that are also up to 16 feet (5 meters) tall, jutting above the savannah like skyscrapers. To allow the termites to survive inside such a large structure exposed to the hot tropical sun, these nests are constructed to provide natural air conditioning (see Chapter 11) (Luscher 1961).

A huge ant mound in northern Minnesota made of soil, probably the home of *Formica montana* (Hymenoptera: Formicidae), being examined by Bill Garnett.

2. Gourmet

People have long prided themselves on exotic or very pleasing culinary dishes. Black truffles add a very pleasing flavor to some dishes, and they are much sought after for this purpose. Truffles are the fruiting bodies of *Tuber melanosporum* Vittadini, a fungus that grows underground in association with trees, especially oaks. Because the fungus grows underground, it is difficult to locate and harvest. However, people found that pigs or dogs could locate truffles by their odor (dimethyl sulfide, Talou et al. 1990).

However, other observant people already knew that flies of the genus *Suilla* (Diptera: Heleomyzidae) feed upon truffles. These people lie on the ground to watch for the flies to show them where truffles lay beneath the soil (Janvier 1963, Laboulbene 1864, Talou et al 1990). These flies like our gourmet treats too!

3. Tunnel Builders

One of the more ambitious tunnel building efforts of all time is currently underway as contractors are starting to build a tunnel for vehicles to cross beneath the English Channel to link England to the European Continent. This engineering feat is not entirely without precedence as many tunnels already carry incredible numbers of vehicles under rivers and estuaries in the United States and Japan. For example, the Holland Tunnel which connects New York to New Jersey was once the longest mechanically ventilated, 3.7 million cubic feet of air per minute, underwater tunnel in the world. It has served as a model for all other tunnels (McKay 1988).

While perhaps not quite as sophisticated, insects are also tunnel-builders, and they have been building their tunnels for a much longer period of time. Embiidina, or web-spinners, live inside tunnels that are formed from silk they have produced, while ants, termites, and soil insects usually construct their tunnels from the material through which they burrow (Von Frisch 1983). In addition to tunneling within wood, some ground dwelling termites also build extension tunnels across inedible stone or concrete to span the distance from their subterranean homes to their next meal. These extensions, which can be several feet in length, protect the termites from desiccation and predation as they move about. Both ants and termites add salivary secretions ("glue") to their tunnels as a binder to harden and strengthen the walls (Holldobler & Wilson 1990). Similar tunnel building is practiced by yellowjackets that nest underground. Not only do they line their tunnels with mud to which saliva has been added, but in the fall they also extend this mud to form "turrets" at the entrance to their nests (Akre et al 1981). Carpenter ants also line earthen tunnels with sawdust.

Other tunnel builders are found among flies (Diptera), butterflies and moths (Lepidoptera). Some members of these groups will tunnel, or mine, through the central tissue (**mesophyll**) of leaves leaving the upper and lower leaf surfaces intact. The serpentine tunnels of these insects can create interesting patterns on the leaves. Early larvae of the horse stomach bot fly (Oestridae) produces a condition in humans called **cutaneous larval migrans** as the larvae tunnel through the upper layers of skin.

An exposed tunnel of the dampwood termite (Isoptera: Hodotermitidae)

9

4. Paper

We take paper for granted. It is used in a vast number of ways-
newspapers, magazines, books, packaging, boxes, and even tableware. Modern
paper is made by machines and is incredibly uniform in thickness and
smoothness, essentially without blemish. However, paper making was once a
laborious process as plant fibers of various types were wetted and spread out to
dry.

Nevertheless, hundreds of thousands of years before mankind's first
efforts, the world's first paper makers were collecting weathered fibers from

plants which were chewed and mixed with saliva to make paper for nests. These paper makers were the yellowjackets (Edwards 1980, McGovern et al. 1988, Spradbery 1973), umbrella or paper wasps, hornets, and the social wasps of the Central and South American tropics, the polybiines (Akre 1982)(all Hymenoptera: Vespidae). The fiber they gather is made into a hard paper, called **carton**, used for constructing cells for their brood and the **envelope** which encloses the nest. The nest envelopes often are of interesting color patterns depending on the source of the original plant fibers. Some wasps even weave bits of sand into their paper for strength and hardness.

The envelope covering most nests is an excellent insulator. It is laid down in layers with air spaces between. Wasps make maximum use of dead air spaces in their nest construction to help regulate the internal temperature of the nest. In addition, some species of yellowjackets build these paper nests in a cavity below the surface of the soil which also tends to protect the nest and brood from fluctuations in temperature.

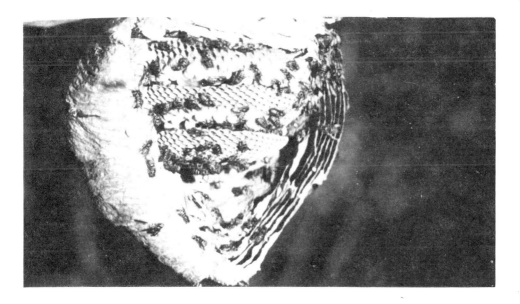

A yellowjacket nest (*Dolichovespula arenaria*,Hymenoptera: Vespidae) with half the envelope cut away showing the layered construction and exposing the combs.

5. Antifreeze

Although antifreezes are used for a number of purposes, the best known is that associated with water-cooled engines. During the winter the radiator and engine block must be protected by an antifreeze which lowers the freezing temperature of the coolant. Without antifreeze the coolant will freeze, expanding enough to damage the radiator and engine block. Most antifreezes contain ethylene glycol or similar substances, although wood alcohol (methanol) has been used.

Insects also have a freezing problem, which they have solved in essentially the same way. When winter arrives many insects enter into a "cold hardy" stage. This can be the eggs, immatures, or adults, depending on insect species. Insects that live through a cold period must resist freezing since ice crystal formation in their tissues can be fatal. The antifreeze used is glycerol, a type of alcohol, produced especially for this purpose. Glycerol lowers the freezing point of the blood (haemolymph) inside the insect and renders it cold hardy (Chapman 1982, Kerkut & Gilbert 1985). In addition to producing antifreeze the clod hardy insects can divest themselves of any excess water to reduce the chance of freezing injury.

A scorpionfly (Mecoptera: Boreidae) on the surface of the snow. These insects are able to move at low temperatures.

6. Camouflage

The need to remain unseen and undetected is critically important to hunters and to the military in times of war. Camouflage netting is used by waterfowl hunters to cover their blinds and by the military to disguise tactical installations. Camouflage is becoming increasingly sophisticated as B-2 bombers utilizing "stealth" technology, which makes them "invisible" to radar, are added to the military arsenal.

Humans learned many of the principles of camouflage and **cryptic coloration** by observing animals, including insects, that use camouflage to their

14

advantage for survival (Cott 1966). Insects make extensive use of color patterns either to match their background (Sargent 1966) or to de-emphasize or disrupt the outline of their body. Often the body outline is further disguised by the presence of brushes or tufts of hair on the edges of the body. This technique is commonly used by cryptic moths and butterflies. Along with disruptive coloration or camouflage, adaptive stillness is also extremely important. The principle, of course, is that movement attracts attention.

One of the most famous examples of cryptic coloration concerns a moth in Great Britain. Initially populations of the English pepper moth, *Biston betularia* (Lepidoptera: Geometridae), were predominately composed of light colored individuals, although a dark form (**morph**) also existed. As soot, generated from burning vast quantities of coal in heavily industrialized areas, accumulated on tree trunks, dark moths were less susceptible to predation by birds and eventually became more common than light colored moths. This was termed **industrial melanism** because the dark, melanistic, morph became more common due to industries effect on the ecosystem. When pollution was brought under control and the tree trunks returned to their natural, light color, survival of the lighter color moths increased until the moth population eventually, once again, consisted mostly of light colored biotypes (Kettlewell 1965, 1973).

Phrictus quinquifasciatus (Homoptera: Fulgoridae) immature. A very cryptic insect.

15

7. Warning Coloration

For humans, red seems to be an international code for danger or warning. Signs warning of danger are nearly always red in color, as are the flags tied to large loads of lumber or pipe that exceed the length of the transporting vehicle.

Red hunting jackets have been used for years as a highly visible warning to other hunters, although blaze orange and other fluorescent colors are used in some areas. Red also seems to have a psychological effect on people and arouses caution or anger.

Insects also use color to their advantage. Many brightly colored insects "intentionally advertise" their presence (Cott 1966, Atkins 1980). Often these insects are distasteful to predators, while some have other defenses such as chemical sprays, stings, or bites. These insects gain an advantage by advertisement and make no effort to hide. They are openly active, "daring" predators to attack. Once a predator attacks and experiences the unwelcome results, it remembers that these brightly colored insects are not good prey. These predators are mostly vertebrates as this bright coloration and distasteful attribute does not seem to work well against insect predators. Another common example of warning coloration is the bright yellow stripes and patterns found on many stinging wasps and their mimics.

A bug (Hemiptera: Scuttelleridae) that is very brightly colored and with contrasting spots. The bright colors warn potential predators of harm that may befall them if they eat this insect,

17

8.   Flight

Only birds, bats, insects, and humans fly. Obviously the first into the air were the insects, and in many ways they are still·the best and most versatile fliers. The basic mechanism of flight is based on a **click mechanism** that suddenly transmits nearly all the stored muscle energy of the thorax to the down or power stroke of the wings (see Chapter 53) (Pringle 1975, Ross et al. 1982, Chapman 1982). Variations in flying abilities include insects that have the ability to hover in one place such as dragonflies (Odonata), bot flies (Diptera: Oestridae), bee flies (Diptera: Bombyliidae) and hover flies (also known as flower flies) (Diptera: Syrphidae). These insects possess a unique flip mechanism which causes the wing to sweep obliquely up and down through a small angle (Weis-Fogh 1975). Certain damselflies (Odonata) in Panama are known locally as helicopter bugs because the movement of the golden **stigma** (pigmented spot on the forewings) as the insect hovers in flight is reminiscent of the colored rotors of helicopters. These damselflies hover in front of spider webs while stealing prey from the web.

18

9. Bikinis

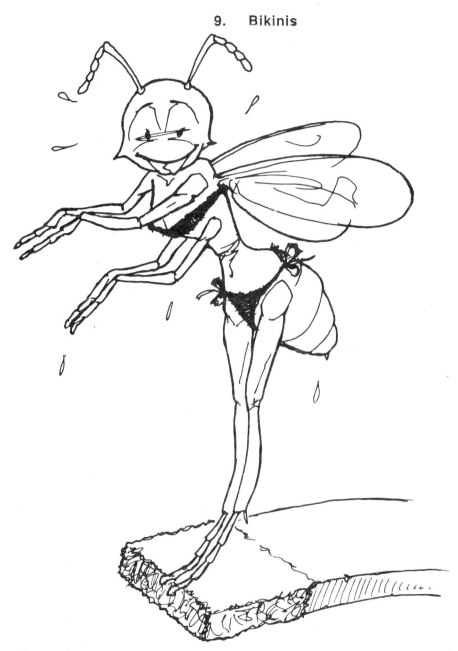

The entomologist who described a distinctive new species of flesh fly (Diptera: Sarcophagidae), *Oxysarcodexia bikini* Dodge, stated that the feminine character was unusually revealed (Dodge 1966). The fly has a fringe of hair on its abdomen that looks remarkably like an abbreviated bikini swimsuit.

10. Communication by Sound

Most of what people learn is seen, not heard. However, sound is obviously important to us as we have spoken languages and appreciate an array of sounds as music. Insects also make music and make extensive use of sound for communication (Wigglesworth 1972, Kerkut & Gilbert 1985). Some have extremely elaborate instruments for the production of these sounds (Snodgrass 1923).

Some termites and ants beat their heads against the walls of their nests to signal alarm, while certain aphids stomp their feet or bang their abdomens on the substrate to signal alarm to other aphids (Bowers 1972). Perhaps the most sophisticated and highly developed use of sound is by bees, including stingless bees and honey bees. The honey bee, *Apis mellifera* L. (Hymenoptera: Apidae), emits pulse trains of sound, produced by wing vibration, during their waggle

20

dance that convey information to nestmates about the distance to a food source (Esch 1967, Gould & Gould 1988, Winston 1987). Species of *Melipona* (Hymenoptera: Meliponidae), a stingless bee, also use a "morse code" of sound to indicate distance to the food source to their hivemates (Esch 1967). Other sounds made by honey bees include the warning buzz of disturbed workers and the piping of the queen that calms disturbed workers. Other queen-produced sounds are tooting (sometimes called piping) and quacking. Tooting is the sound made by a virgin queen soon after she emerges as an adult, and quacking is the sound made by new queens that are forcefully retained inside their cells by the workers. Eventually, they are released to challenge all other queens (Wenner 1964).

The most complex song known for insects is produced by Uhler's katydid (Orthoptera: Tettigoniidae)(Walker & Dew 1972), but even the lowly fruit flies (Diptera: Drosophilidae) produce songs of love and courtship (Bennet-Clark & Ewing 1970).

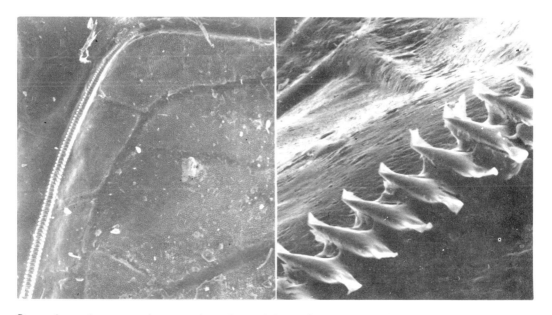

Scanning electron micrographs of a cricket (Orthoptera: Gryllidae) wing showing the file used for sound production magnified about 30 (left) and 400 (right) times.

On a hot summer day an air conditioned house is a welcome relief. The air is cooled, filtered, and blown into the interior by elaborate machines. However, termites accomplished essentially the same type of air conditioning hundreds of thousands of years before humans. The most sophisticated system yet investigated is used by the African termite *Macrotermes natalensis*. Nests of this insect are constructed of soil molded into hollow mounds that can be 16 feet tall by 16 feet in diameter at the base. The walls are 16-23 inches thick, and have numerous ridges on the outside which contain tunnels for circulation and diffusion of gases through the walls.

Termites are extremely susceptible to dryness and will die quickly when exposed to dry air. Their nests are moist, usually with internal humidity ranging from 98-99%. Temperature in the center of the nest rarely varies more than a few degrees. The content of carbon dioxide in the air at the center of the nest is also regulated to 2.7%, a relatively low figure considering the number of individuals living and breathing in this closed space. A medium sized colony of about 2 million individuals needs 240 liters of oxygen a day.

The supply of oxygen and regulation of temperature and humidity are accomplished by the design of the nest. The center of the nest contains most of the workers and the fungus grown (see Chapter 17) by the colony for food. As the air in the center heats it rises to a hollow at the top known as the attic. It then travels down the tubes in the ridges on the outside surface of the nest. As it moves through the tubes carbon dioxide is replaced with fresh oxygen, and the air is cooled. Water is brought up from deep in the soil under the nest to maintain the high humidity. Food brought into the nest also contributes to the water supply (Luscher 1961)

Honey bees also used forced air evaporative cooling to regulate the internal temperature of the hive. Worker bees will align themselves at the hive opening fanning their wings to ventilate the hive.

12. Food Storage

Food storage systems have been used by people for thousands of years. In many ways the development of effective food storage systems was an important element for the success and spread of our civilizations. Today we have elaborate systems for storage, canning, refrigeration, and freezing to preserve foods. However, insects were the first to store food. Honey cask ants (Hymenoptera: Formicidae, genus *Myrmecocystus*) of the Southwest have special workers, called **repletes**, which are used by the colony as living storage containers for "honey"

gathered by other workers. The colony members can then beg for food from the repletes during times of need (Wheeler 1910, Wilson 1971). Native Americans of the Southwest also gathered and ate the repletes. Surprisingly, the fire ants introduced into the southern United States also have repletes (Glancey et al. 1973, Lofgren & Vander Meer 1986) as do species of carpenter ants (Wilson 1971).

Ants are not the only insects to store honey. All species of honey bees and most other social bees store honey. Honey also is stored by polybiine wasps (Richards 1978) and by paper wasps (genus *Polistes*) (Strassman 1979).

Harvester ants collect and store seeds for food. The best known of these ants are species of *Pogonomyrmex,* common to the southern and western U.S. (Wheeler 1910, Wilson 1971). (Also see Chapter 65)

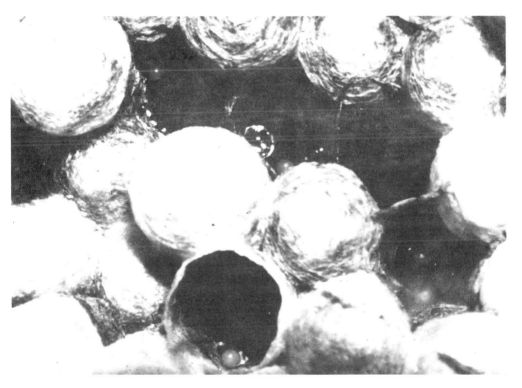

Some wasps and bees make hexagonal cells which are used to hold developing young and store food.

13. Domestic Animals

Humans have domesticated a few species of animals to use as beast of burden or as food. The more common domestic animals are cows, horses, pigs, chickens, dogs, and cats. However insects kept "domesticated animals" much earlier than we did. One such association is between certain ants and the larvae of lycaenid butterflies (blues) (Lepidoptera: Lycaenidae). The caterpillars possess special glands, called **Hinton's glands**, which produce a chemical which attracts ants. They also have a honey or **Newcomer's gland** which produces a substance the ants like to eat (Kistner 1982). In return, the caterpillars are protected from insect predators by the ants.

26

A more familiar situation is presented by ants tending aphids (Homoptera: Aphididae) and other Homoptera (Wilson 1971). This relationship is called **mutualism** because both the ants and the aphids gain mutual benefit from the association. Mutualistic relationships exist between many genera and species of ants. Not only do ants carry aphids to their host plants and protect them (Wilson 1971), but at least one ant species responds aggressively to an alarm pheromone (see Chapter 23) released by the aphids. When the aphids are disturbed the ants will attack any insects or other intruders in the immediate area (Kistner 1982).

Ants, as well as some other kinds of insects, collect a sugary substance called **honeydew** which is excreted by a number of other sap-feeding insects in the Order Homoptera including scales (Coccidae), pine/spruce aphids (Chermidae), psyllids (Psyllidae), treehoppers (Membracidae), leafhoppers (Jassidae), froghoppers (Cercopidae), and planthoppers (Fulgoridae) (Wilson 1971). In one species, *Trabutina mannipara* (Ehrenberg) (Coccidae), the material is so abundant at times that it is collected by both ants and humans for food. This is believed to be the manna mentioned in the Bible.

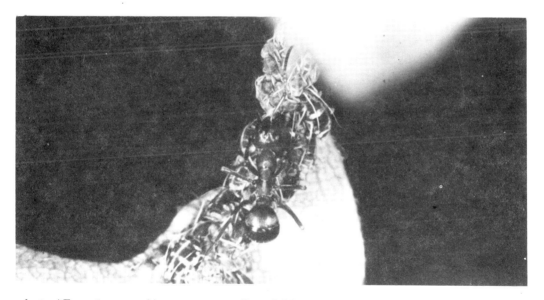

Ants (*Formica* sp., Hymenoptera: Formicidae) tending aphids from which they will collect honeydew.

14. Highways

It might seem somewhat ludicrous to draw an analogy between the superhighways of human civilizations and primitive trails made by insects, but the two serve much the same function. In many cases, the trails made by insects are not simple but are really quite complex.

Among the most noticeable of all insect trails are those of tropical army ants (Rettenmeyer 1963, Schneirla 1971). These consist of unbroken lines of ants, often with ants traveling in both directions. Army ants are blind, they rely completely on their own chemical trails for their orientation. These ants follow trails marked by fellow workers, and often reenforce the trail as they travel by depositing a chemical substance of their own.

Army ants are carnivorous, and use chemical trails to find and raid other insects. The best known species are also nomadic and move (emigrate) distances of up to a hundred yards daily before they establish a new bivouac or nest. They usually emigrate over a strong raiding trail established during the day.

Although not nearly as noticeable, army ants also occur in the United States as far north as Sioux City, Iowa. Like their tropical relatives, these ants are blind and rely solely on chemical trails for foraging and emigrating.

Many other ants use chemical trails and "highways." The broad trails of leafcutting ants are an extremely noticeable feature of many areas in subtropical and tropical areas. These trails, up to 4-6 inches wide in some areas, are carefully maintained by workers that clear away all obstructions, including vegetation, so that the trails are totally bare and well groomed. Therefore these trails are both physical and chemical. Workers use these trails to forage for leaves that the workers cut and carry back by holding the pieces above their bodies. Because of this habit they are sometimes called parasol ants (Weber 1972, Wilson 1971).

Although it is not nearly as well known, carpenter ants also establish trails similar to those described for leafcutters. In the Pacific Northwest it is not unusual to find trails 1-1.5 inches wide cut through well manicured lawns.

Since these ants do most of their foraging at night, these trails are chemical as well as physical routes(Hansen & Akre 1985).

Trail of a carpenter ant (*Camponotus modoc*, Hymenoptera: Formicidae) near Port Angeles, WA. The ants have removed all debris from the trail to make travel easier. The trail is also marked with a chemical trail pheromone.

15. Recycling

Some of our largest recycling efforts have been during times of war when raw materials were at a premium. Scrap drives for paper and metal were common durin World War II. Recycling also has become a constant and expanding practice in peace time. We have entered a period of environmental concern, in which we have realized that we must recycle our waste paper, plastic, and metal or be suffocated in them.

Insects have been recycling for millions of years. Among the more obvious of the recyclers are the honey bees. Beeswax combs are used repeatedly in the yearly storage of honey, pollen, or brood. Even the wax, used to construct the combs, is reused within a hive.

Slavery of humans by others, unfortunately, still occurs in our imperfect human world. Even though in most enlightened civilizations slavery has been abandoned, in some instances, it has merely been replaced by various forms of economic slavery. Either way the goal was, and is, to force others to do work at little cost to the masters.

Slavery is common among ants with at least 35 species that depend on slave labor for their survival (Wilson 1975). Indeed, most slave-making ant species are so specialized that they will starve if deprived of their slaves. Perhaps the best known of all slave-making ants are several species in the genus *Polyergus* (Topoff 1984). These ants have specialized sickle-shaped, sharply pointed mandibles that are excellent for piercing the bodies of other ants, but poorly suited to other tasks. As a result they cannot feed themselves. In addition, the *Polyergus* spray the raided colony with a chemical, called a propaganda pheromone, that causes the victimized ants to abandon their nest and scatter in all directions. When this occurs the slave makers simply gather up pupae of the host ant and return to their nest. A few of the raided ants will try to fight, but they are no match for the formidable raiders.

The normal behavior of slave making ants consists of periodically raiding colonies of closely related ant species (various species of *Formica*) to capture pupae that they bring back to the nest. Some of these pupae are used as food, but most are reared to adults in the colony. When these slaves emerge as adult ants, they are integrated into the colony and perform all the normal colony functions such as feeding the brood and adults, cleaning the nest, and even taking part in the defense of the colony. The *Polyergus* workers do little except beg from their slaves and groom themselves.

Similar strategies have evolved in social wasps, but in this case the parasitic species have advanced to a point where they produce no workers of their own, but rely on the host workers to rear their brood (Akre 1982). Highly evolved parasitic ants exhibit similar life styles.

33

17. Gardening

The original gardeners were insects. Leafcutting ants collect plant materials to use as a medium for growing fungus to be used as food for themselves and their brood (Weber 1972, Wheeler 1910, 1973, Wilson 1971). Some termites of Africa and Asia also grow fungus for food (Batra & Batra 1967). Among these termites are species of *Macrotermes* which have large air conditioned nests (see Chapter 11).

The fungus ciltivated by insects is able to utilize **cellulose**, a major component of plant tissues. Most insects, including the gardening insects, can digest cellulose only with the help of specialized microorganisms, called symbionts, which live in the gut of some insects. Therefore by growing fungus the insects are able to take advantage of a source of food, the cellulose, which they could not normally use. The fungus is frequently a single species kept pure by secretions from the gardeners (see Chapter 26). The fungal mass can consist of small pockets found throughout the nest or it may be a single large ball. One of these balls found in a termite nest was two feet in diameter and weighed 60 pounds (Batra & Batra 1967).

Often ants and termites are extremely destructive of farm crops and structures while gathering material to grow fungus. In Central and South America leafcutting ants, usually *Atta cephalotes* (L.), can cause extensive defoliation to citrus trees (Weber 1972). Equally well known is the destruction of all kinds of crops by termites in Africa and Asia. Some fungus growing termites will fertilize their fungus crop with their own excrement.

Trail of a leafcuting ant in Panama with many of the nest-bound workers carrying pieces of leaves to be used for growing fungus on which the ants feed.

Sailors, who need to orient in a certain fixed direction in the absence of landmarks, rely heavily on celestial bodies, such as the sun and stars, for direction. Many insects, particularly ants (Wehner 1976) and honey bees (Von Frisch 1966, 1967) also use the sun for orientation. Honey bees are able to travel at a fixed angle to the sun to return to the hive. Once there, they are able to convey these directions to hivemates inside the darkened hive. They do this through a "dance" which gives directions to guide others to the food source. Incredibly they even compensate for changes in the sun's position through the day. However, the sun cannot be relied upon during heavily overcast days when it is obscured and as its position in the sky is difficult to determine (see Chapter 62). At such times the bees use a compass.

Natural magnets, such as lodestones (magnetite or Fe_3O_4), that could be used to point to magnetic north were prized by early human travelers. These primitive compasses evolved into the compasses commonly used today. Magnetic navigation systems have been used by other organisms for eons. Even certain bacteria are magnetotactic and tend to swim along magnetic field lines influenced by the magnetic field of the earth (Blakemore & Frankel 1981). Their compass consists of magnetite (lodestone) which they synthesize from soluble iron. More sophisticated uses of the magnetic field of the earth for orientation are probably used by honey bees. Honey bees can detect and will orientate their dances to the earth's magnetism (Winston 1987, Dyer & Gould 1983, Gould & Gould 1988). Their compasses may consist of a region of transversely oriented material in the front of the **gaster** (abdomen behind the **pedicel**, or waist) and of bands of cells in each abdominal segment that contain iron granules.

19. Sewing or Lashing

It was a real achievement when humans devised methods to attach materials securely together with pliant strands of plant fiber or sinew to form shelters and clothing. Weaver ants of Africa and tropical Asia employed primitive lashing and securing techniques long before humans did (Holldobler & Wilson 1977, 1983). Most weaver ants are species of *Oecophylla* that build nests by folding leaves into tentlike compartments secured by silk "lashing ropes" spun by their larvae. When several workers are able to pull two leaves together, other workers use larvae like shuttles to spin silk across the gap. Thousands of these silken threads are needed to hold the leaves in place. Most remarkable is the cooperation between these insects in pulling the leaves together. When leaves are too far apart the workers form living bridges between the leaves by hanging onto one another. Additional workers then crawl over the backs of these workers while holding the edge of the leaf to shorten the gap. After many of these shortenings, the workers then use larval silk to hold the leaves in place.

Tent caterpillars (Lepidoptera: Lasiocampidae) will form shelters by lashing leaves together with their silken webbing.

20. Scuba Diving

It seems that human curiosity cannot be satisfied. We are always driven to explore even the most inaccessible regions of our world. Since very early times people have attempted to carry breathable air with them below the surface of the water. Some of the first attempts were rather crude with animal skins being used to hold a limited supply. This enabled individuals to go down a bit farther or stay down a little longer. The scuba diving gear of today is the outgrowth of these early trials, but is much more sophisticated, allowing us to reach great depths and to remain submerged for a long time.

Insects have also employed "scuba diving" techniques. Some make use of air bubble breathing in which the insect at the water surface "grabs" a bubble of air to carry with it as it dives. The oxygen in the bubble is gradually depleted, eventually forcing the insect to return to the surface for a new bubble. Others employ a **plastron** (Gillott 1980, Romoser 1981), a more elaborate form of respiration. A plastron is a thin layer of gas usually held in place against the insects breathing holes, or **spiracles**, by specialized (**hydrofuge**) hairs or other modifications to the cuticle. The plastron then acts, not as a supply of stored oxygen, but as a gill with oxygen diffusing into the plastron as it is used. Insects with this type of respiration usually do not have to restore the oxygen by visits to the surface. Thus they can stay submerged indefinitely. Both types of respiration are used by aquatic bugs (Hemiptera) and beetles (Coleoptera).

A backswimmer (Hemiptera: Notonectidae) at the surface of the water.

41

21. Chemical Defense

Insects are masters at the use of noxious or stinging (**urticating**) chemicals for their defense (Atkins 1980, Bell & Carde 1984, Blum 1981, Roth & Eisner 1962,). One example is the spraying of **formic acid** by ants. Indeed, the family name, Formicidae, is derived from this attribute, although many ants do not produce this acid. A colony of thatching ants (*Formica* spp), will produce a gaseous/liquid pall of formic acid if disturbed. This pungent material, which "takes the breath" of most mammals, is also effective against many invertebrates such as ants, beetles, and millipeds. Indeed, some birds even use thatching ants to rid themselves of lice and mites. This is called **anting.** Even people have placed their furs and blankets on ant mounds for the workers to remove lice and fleas. The disturbance of the colony also probably triggered many ant workers to "shoot" formic acid onto the fur which would also help to rid it of vermin.

Another example of chemical defense is that of the bombardier beetle, *Brachinus* (Carabidae), that produces a hot (100°C) spray of chemicals, called **quinones**, when disturbed or attacked (see Chapter 74). Quinones are so reactive that they exist in the beetles only as separated precursors that are mixed immediately prior to use. **Benzoquinones** are the most widely found group of all the defensive chemicals although many other materials are used such as aldehydes, ketones, and even steroids. **Steroids**, which are produced only by diving beetles (Dytiscidae), are among the most complex of all defensive chemicals. When these beetles are swallowed by fish, the steroids cause the fish to regurgitate the unharmed beetle (Schildknecht 1970, 1971).

Another familiar example of insect chemical defenses are blister beetles (Meloidae). When these beetles are carelessly handled they secrete a fluid which quickly raises a blister on human skin. Spanish fly, believed by many to act as an **aphrodisiac**, is derived from a blister beetles, *Lytta vesicatoria* (L.). Spanish fly contains a powerful toxin that can cause serious damage or even death to humans (see Chapter 22).

Perhaps one of the most unusual uses of a chemical defense is the use of a spray by female ground (Carabidae) beetles against unwanted attention by arduous males. The spray physically incapacitates the male so the female can escape (Kirk & Dupraz 1972). Thus, we have the first use of "Mace" by an insect.

Larva of an alder leaf beetle (Coleoptera: Chrysomelidae) releasing a defensive secretion from glands along the sides of its abdominal segments.

Insects, just like humans, have trouble with microbes. Diving water beetles (Dytiscidae) produce an antibiotic paste to keep their body surface microorganism-free. This helps keep the surface smooth for friction-free swimming, and reduces the number of invasive bacteria with which they must contend. The components of the paste are **benzoic acid** and several phenols, particularly **methyl p-hydroxybenzoate** and **p-hydroxybenz-aldehyde**, and a glycoprotein. The paste is excreted from special glands and spread onto their bodies with their hind legs while the beetles are sitting on emergent vegetation. IThe paste hardens, killing and encapsulating many microbes, and is washed off when the beetle returns into the water (Schildknecht 1970, 1971). These materials, particularly the methyl p-hydroxybenzoate and similar compounds, are used by humans to eliminate microbes from canned foods.

Insects antibiotics are used not only on themselves. Insects and their products have also been exploited for medicinal purposes by humans. People have attributed curative powers to insects for hundreds of years. In the 1600's these powers were ascribed to nearly every insect (Metcalf et al. 1962). However, it is now known that certain insects do indeed have special medicinal properties. Perhaps the most famous case was recorded by Dr. W. S. Baer. While observing the wounds of soldiers in WW I, he noticed that injuries infested with green bottle fly maggots (Diptera: Calliphoridae) did not develop gangrene whereas those treated and sometimes even dressed promptly frequently did. Further investigations showed that the maggots ate only putrid flesh, and that they were much more efficient than physicians in the cleaning of deep wounds. Because of this blow flies were sometimes reared under sterile conditions to be used specifically for this purpose, and in rare cases are even used today. An additional benefit was discovered in 1935 when **allantoin**, a substance that promotes healing, was found to be excreted by the maggots. Of course, mostly synthetic allantoin is used today (Metcalf et al. 1962).

Another medicine produced by insects is cantharidin, the defensive chemical released by blister beetles. A particularly well known insect in this group is *Lytta vesicatoria* (L.), the Spanish fly (see Chapter 21). Extracts from this beetle were used as an aphrodisiac for humans until it was realized that this material is extremely harmful. It is still used today for the treatment of diseases of the urogenital tract, and for the breeding of animals.

The venom from the sting of honey bees is probably the most wide spread insect "medicine" in the world. Bee stings are used by many people to treat their arthritis and many other ailments. Royal jelly, another honey bee product, is also used by many, but it has no known medicinal value.

45

23. Perfumes/Pheromones

A **pheromone** is a chemical that is secreted by one individual and causes a physiological or behavioral response in another individual of the same species. Usually pheromones are divided into two groups depending upon the response elicited. Physiological pheromones are called **primers**, and they cause a long term effect. A well known example is the queen pheromone of honey bees that, among other things, prevents the production of new queens in a "queen right"

colony. Other pheromones, called **releasers**, cause an immediate behavioral response in the receiving individual. An example, using the honey bee again, is the alarm pheromone, coating the bees stinger. This odor attracts and excites other bees to sting near the same site (Gould & Gould 1988, Von Frisch 1967, Winston 1987). Other pheromones commonly used by insects are sex pheromones used to attract individuals of the opposite sex for the purpose of mating (Jacobson 1971, 1972). These chemical attractants are produced by females, by males, or in some cases, by both (Beroza 1971).

The investigations of pheromones essentially started with insects and continues at a fantastic pace because insects are masters of the uses of chemical signals (Wilson 1963). Even though insects have been the subjects of most pheromone research (Birch 1974, Jacobson 1972, Shorey 1973, 1976), it is now known that other animals, including humans, also produce pheromones. However, natural human pheromones have been largely replaced by artificial pheromones such as perfumes, colognes, and after shaves.

Worker carpenter ant (*Camponotus* sp., Hymenoptera: Formicidae) in a typical defensive stance with the gaster curled under and foward so the ant can spray formic acid toward an antagonist.

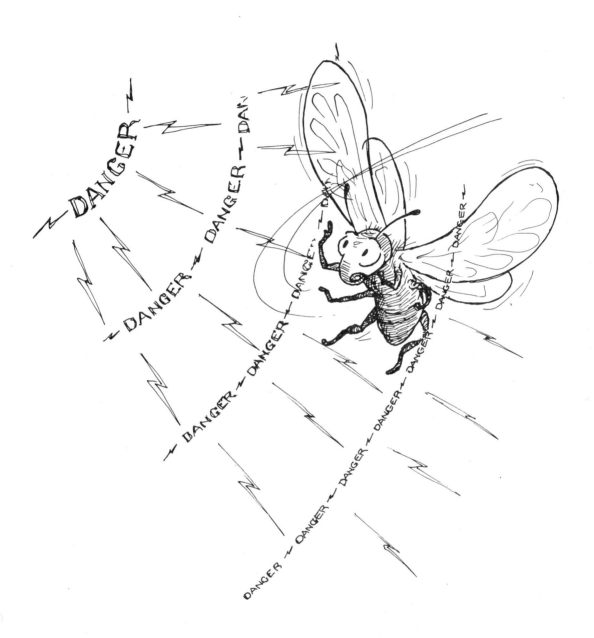

48

The ability to detect airplanes through the use of radio waves that bounced off a target was an technological achievement developed during the World War II. A similar principal using sound waves was then applied to submarine detection resulting in the development of sonar. Today we have progressed far beyond these early accomplishments. Even sport fishermen use inexpensive, but sophisticated sonar devices to locate fish; some devices will even indicate depth and size of the fishes.

Insects also make use of both radar and sonar. Moths of the families Arctiidae, Geometridae, and Noctuidae can detect the chirps of echo-locating bats (Roeder & Treat 1961, Roeder 1965, Dunning & Roeder 1965). When moths are exposed to bat radar they instantly take evasive action to avoid being eaten. They fly in erratic loops, take sharp dives, or simply fly away at top speed. Moths that are deaf because their "ears" or **tympana** are incapacitated are not able to detect the bat's radar and are quickly eaten. At least one species of green lacewing (Neuroptera: Chrysopidae), *Chrysopa carnea* Stephens, has tympanal organs which are also used to avoid predation by echo locating bats (Miller & MacLeod 1966).

Sound production in aquatic insects can be compared to sonar in caddisflies (Trichoptera), dragonflies (Odonata), beetles (Coleoptera), and some water bugs (Hemiptera) (Aiken 1985). This sound has various functions, but the most common are defense and communication. For example, whirligig beetles (Gyrinidae), can detect and avoid each other, and locate prey by using the waves they generate while moving (Tucker 1969). The waves bounce back to them, allowing them to detect the position of objects on the water surface. This use of vibrations (waves) is similar to sonar.

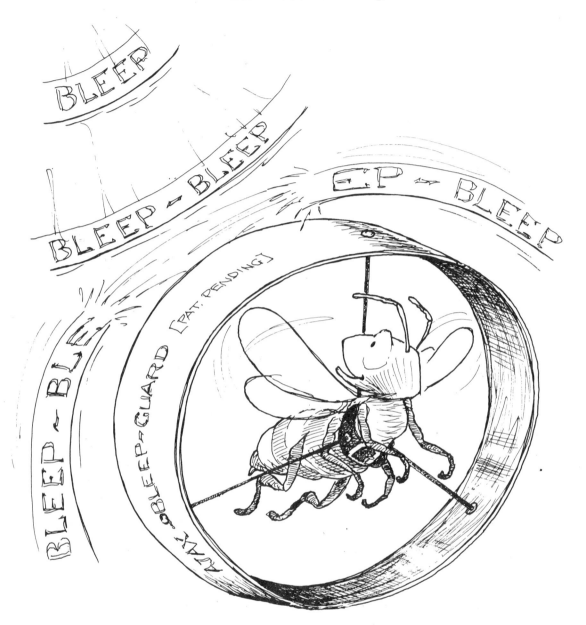

The echo-location sounds emitted by bats are not only received by arctiid moths, but the moths emit trains of clicks that are used to jam the radar of the bats thus helping the moth to escape predation (Dunning & Roeder 1965).

26. Fungicides

Today the most frequently applied agricultural chemicals in the United States are herbicides used to control weeds in our crops. Insects also have weed problems. The leafcutting ant, *Atta sexdens* L., needs to control "weeds" in the form of unwanted fungi in its food crop, the fungus on which the colony feeds (see Chapter 17). These fungi could be a real problem to the ants, without the use of a fungicide, **myrmicacin** (beta-hydroxy-decanoic acid), which is produced in their **metathoracic glands**. The myrmicacin does not affect the beneficial fungus in their culture so it can be used to keep their "crops" free of weeds (Schildknecht 1971).

Humans tend to think of themselves as the only truly social animals with rules, conventions, and life styles. However, many animals are social, and often establish elaborate dominance hierarchies that govern their behavior (Gould 1982). This ensures that the individuals at the top of the social structure are always the first for food, water, and mating privileges. In somewhat modified fashion, according to our social customs, these same rules govern human behavior.

Truly social, or **eusocial**, insects are defined as those having a reproductive division of labor, an overlap of generations, and cooperation in caring for the young. Eusociality has evolved at least 13 times in insects (Wilson 1971, Michener 1970, for a comparison with a primitive human society see Hill & Hurtado 1989). Social insects include all the ants and termites, and many of the bees and wasps. One recently discovered instance of sociality in an insect was of a small sphecid wasp, *Microstigmus comes* Krombein, that occurs in Costa Rica (Matthews 1968). However, students of bee behavior are uncovering eusocial behavior in new groups at a rapid pace.

Carpenter ant (*Camponotus* sp. Hymenoptera: Formicidae) major and minor workers attending to the needs of their larvae. Ants are eusocial insects with distinct division of labor.

28. Glue

Glues have an important role in our daily lives. Early glues, derived from plant and animal sources, were often frustratingly inadequate. Today incredibly powerful and specialized glues are capable of adhering to almost any kind of material. In addition to more mundane uses, glues are used to join human tissues during surgery, and even to attach ceramic tiles to the space shuttle.

Insects also use glues. The most common use of glues by insects is to affix their eggs to the substrate so that they cannot be displaced easily. Egg glueing takes many forms. The bright orange eggs of the Colorado potato beetle are glued into place in small clusters on a potato plant, and the eggs of lacewings are individually suspended on long stalks glued to the surface of a leaf. Animal parasites, such as lice, often glue their eggs to the pelage (hair, feathers) of the host. These glues are produced by **accessory glands**, known as collaterial glands, associated with the insects reproductive tract (Romoser 1981).

During 1990 the popular press reported that a construction engineer in Brazil had synthesized the salivary substances used by termites to glue their earthen nests together and that this material was being used to build roads in the tropics. The substance was rumored to be as good or better than asphalt, and much cheaper.

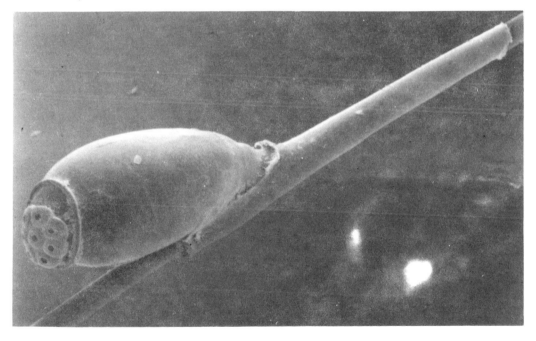

A scanning electron micrograph of an egg of a human head louse (*Pediculus humanus capitis*, Phthiraptera: Pediculidae) securely glued to the shaft of a hair magnified about 100 times.

29. Dominance Hierarchies

A number of animals exhibit dominance hierarchies in which the dominant individual is first in nearly all social interactions including feeding and mating. Chickens have well known pecking orders in which the alpha or top hen has the right to peck all subordinates and not be pecked in return. Cows have well established butting orders, and the dominant cow is always in the lead when they return to the barn with the other cows following according to rank. Dominance hierarchies also exist in troops of baboons, in monkeys, and in packs of wolves. Very similar types of social interactions occur in humans. In many animals dominance is settled and maintained by intimidation and sometimes combat. Often physical combat has been replaced with less damaging contests, and the subordinate always assumes a submissive posture. A familiar submissive posture which has been seen by most people is a young puppy turning on its back to expose vunerable areas to a dominant dog (or to a scolding human) and urinating.

Insects also establish and maintain dominance hierarchies. One of the first reported was for bumble bees (Huber 1802). A better known example concerns a paper wasp, *Polistes gallicus* (L.). A scientist studying the division of labor among females realized that the behavior he was observing was similar to the dominance hierarchies previously reported for other animals (Pardi 1948) He discovered that, while several females may cooperate in building a nest, one is the dominant ("queen") and as such lays most of the eggs, while the others assume the duties of workers. Cooperation enables colonies to survive that probably would perish if only a single female tried to establish the nest (West 1967). Female ants also establish dominance hierarchies (Heinze 1990).

Field crickets establish hierarchies associated with territoriality (Kato & Hyasaka 1958) as do dragonflies (Odonata) (Jacobs 1955, Ito 1960), bot flies (Diptera:Oestridae) (Catts 1967), and other insects. *Myrmecophila manni* Schimmer (Orthoptera: Gryllidae), a tiny cricket found only in association with ants establishes linear hierarchies by fighting (Henderson & Akre 1986). Cockroaches also have dominant-subordinate relationships that are established by butting and pushing. Even though the subordinate is apparently not hurt, many die, presumably from stress (Ewing 1967). This non-specific stress is also very important in human health (Selye 1973), but, as usual, the insects were first with this dubious honor.

57

Insects are unquestionably their own worst enemies. If there were not beneficial species as parasites or predators of insect pests on crop plants there would be substantially less food available for human consumption around the world. Foremost among the predaceous insects are the ants, considered by many to be the ecological dominants of the terrestrial world (Gotwald 1986, Holldobler & Wilson 1990). Ants were the first insects used by humans as biological agents to control insects. Weaver ants were used to control citrus pests in China more than a thousand years ago (Liu 1939). Insects that prey upon other insects include beetles, bugs, flies, wasps, dragonflies, lacewings, carnivorous grasshoppers, and even a few caterpillars (Ross et al. 1982, Borror et al. 1989, Metcalf et al. 1962, Mackerras 1970). In addition to predators, there are many species of parasitic insects that kill their hosts. These are correctly called **parasitoids**, to distinguish them from parasites which do not normally kill their hosts. There are many species of parasitoids, but most are tiny wasps (Hymenoptera) or flies (Diptera). While all these examples concern insect control as they benefit mankind, the insects themselves also benefit from these interactions which are essential to maintain the ecological balance. The well being of herbivores, predators/parasites, and their habitat are all linked in the the complex fabric of life.

Scanning electron micrograph of a parasitoid wasp (Hymenoptera: Encyrtidae) emorging from an immature pear psylla (*Cacopsylla pyricola*, Homoptera: Psyllidae) magnified about 60 times.

Gift giving in return for favors is an old human custom also practiced by insects. Female insects usually have a great aversion to being touched, probably because contacts are usually by predators looking for a meal. When the male approaches a female for mating, he frequently must overcome this aversion. The female must also be receptive to mating. In the case of balloon flies (Diptera: Empididae), the courtship ritual has become very elaborate (Kessel 1955). Some male balloon flies simply present food, such as a captured insect, to the female so she can feed as they mate. Others wrap their gift by spinning a silken cocoon around the prey before presenting it to the female. Still others do not capture prey at all, but merely spin a large empty balloon which is presented to the female to occupy her as mating takes place. These balloons are white and attract females from some distance. They are much larger than a normal prey item and are thus a "overoptimal or supernormal stimulus" for mating.

Another gift giver is a bug (Hemiptera: Lygaeidae) of the genus *Stilbocoris*. Males of *Stilbocoris natalensis* Stål gather a fig seed for their prospective mate, and inject saliva into the seed to predigest it for the female. Males without seeds are not successful in mating (Thornhill & Alcock 1983). Gift giving also occurs in two genera of scorpionflies (Mecoptera) (Engelmann 1970).

Few insects actually produce their own building materials, but honey bees are an exception. Workers possess **hypodermal glands** on the bottom of their abdominal plates (sterna) that produce bees wax used for construction of the combs in the hive. Workers consume about 8 pounds of honey to produce 1 pound of bees wax. The combs are the 6 sided cells that hold eggs and developing brood, and also serve as vats to store honey and pollen (von Frisch 1967).

62

Insects also manufacture and secrete silk for construction. Among these silk manufacturers are the Embiidina, or web-spinners, that have a silk gland inside the basal foot segment (**basitarsus**) of the front legs. The silks are used to line the tunnels in which they live (see Chapter 3) (Borror et al. 1989). Many moths and other insects produce silk which is used to construct silken cocoons in which they pupate (Mackerras 1970). The silk of certain species of moths has been used as a source of silk to weave fabrics. The best known of these moths is *Bombyx mori* L, the silk moth. Even some species of aquatic Diptera, the blackflies or simuliids, which seem to be unlikely silk producers, nevertheless produce silk for underwater cocoons (Borror et al. 1989, Kim & Merritt 1987).

Honey bee (*Apis mellifera*, Hymenoptera: Apidae)worker showing the scales of wax produced by the abdominal hypodermal glands. One wax scale has been removed to show size.

33. Plywood

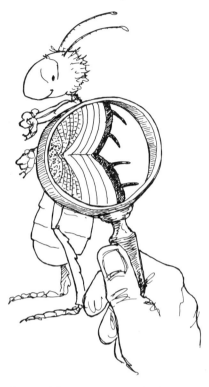

We peel thin sheets of wood from trees and glue them together to form plywood. The resulting multiple layer composite has great strength. Similar laminar construction is used in the deposition of the insect integument or **exoskeleton** (Romoser 1981, Kerkut & Gilbert 1985). The integument is comprised of one cellular layer, the **epidermis,** and two acellular layers, the **cuticle** and the **basement membrane.** The cuticle portion of the integument also consists of a number of different layers. One of these, the **endocuticle**, is comprised of growth layers aligned at different angles to one another much like the layers of wood in plywood. The layers are deposited differently depending on whether they are deposited at night or during the day. **Mirofibrils** within successive sheets deposited during the day are oriented in one direction, these layers do not have a layered (lamellar) appearance. Those deposited at night are oriented in different directions giving them a lamellar appearance.

34. Brain Washers

With the advent of the cold war and torture of political prisoners, the term "brain washing" became common in American slang. It referred, of course, to manipulating the minds of people by subjecting them to extreme degrading duress or torture.

Insects undergo actual "brain washing" throughout their lives as a normal occurrence. They have an open circulatory system with a delicate membrane called the **dorsal vessel** lying along their back, mostly in the abdomen, that pumps their **hemolymph**, or blood, forward. This vessel is composed of two sections the posterior heart and the anterior aorta. The **aorta** expels blood directly at the base of the insects brain, flushing it with nutrients and hormones(Gillott 1980). Hence we have the original case of "brain washing."

65

In the 1950's and 60's many paper mills used caustic chemicals for digestion, bleaching, and blending wood pulp. These chemicals were corroding to the iron or steel pipes that conducted raw materials to the blending tanks and paper machines. One solution to the corrosion problem was the use of tongue and groove oak strips to line the pipes preventing the mix from directly contacting the metal pipes.

Some insects have a similar problem. Their food can cause physical abrasion of the lining of their alimentary tract. To prevent such damage their gut is protected by an intricate, chitinous lining, called the **peritrophic membrane**. The peritrophic membrane is produced from either, epithelial cells at the anterior end of the gut near the cardiac valve, or delaminated from the epithelium of the entire wall of the gut. The membrane protects the gut from physical abrasion, and it is differentially permeable so that only molecules of a certain size can pass through to be absorbed for use by the insect. The indigestible, rough portion of the food continues out the hindgut, still enclosed by the peritrophic membrane, and is excreted as "packaged waste". Because the peritrophic membrane passes out of the body with the waste, it must be produced continually by the insect to maintain protection (Chapman 1982, Kerkut & Gilbert 1985, Romoser 1981, Ross et al. 1982, Richards & Davies 1977).

This memebrane can also serve another function. Larvae of *Ptinus tectus* Boieldieu, a spider beetle (Coleoptera: Ptinidae), make their cocoons from the unbroken peritrophic membrane extruded from the anus (Richards & Davies 1977).

67

In many human societies, but especially obvious in India, there is a division of people into **castes** by social status and function. A member of one caste cannot aspire to rise in society and cannot marry outside class lines. Social insects also have definite female castes called queens and workers. The queen's role in society is to mate with the males (sometimes called **drones**) and produce offspring. The queen in social insects colonies can be compared to a human dictator with her attendant amazons, as the workers are all females in bee, wasp, and ant colonies. The workers may be further subdivided based on their physical attributes. Soldiers may have large heads and mandibles used in fighting, minors are small workers better suited for work within the nest. These castes are rigidly controlled by chemicals called pheromones. The effect of the pheromones is modified somewhat by the nutrition of the individuals as they develop (Holldobler & Wilson 1990, Wilson 1971).

These are all the same species of carpenter ant (*Camponotus modoc*, Hymenoptera: Formicidae), the difference in appearance is due to sex and caste. Clockwise from the top right: winged (**alate**) queen, winged male (notice the size of the head compared to the thorax), major worker, medium worker, minor worker.

69

SPIRACLES

The radiator hose connected to most water cooled engines is reinforced with a spiral or coil of steel to prevent the hose from collapsing under pressure as the coolant in the radiator expands and contracts. Similar reinforcement occurs in the insect's respiratory system. An insect's respiratory system is made up of an extensive, continuous network of air tubes (**tracheae**) that branch until each of the hundreds of thousands of cells in the body is reached by a minute air tube (**tracheole**). The linings of the trachea are made of the same tough material as the exoskeleton. In addition these tracheae have spiral thickenings of the intima, or lining, called **taenidia** to prevent collapsing. This gives them an appearance very similar to an engine radiator hose (Borror et al. 1989, Chapman 1982, Romoser 1981, Ross et al. 1982).

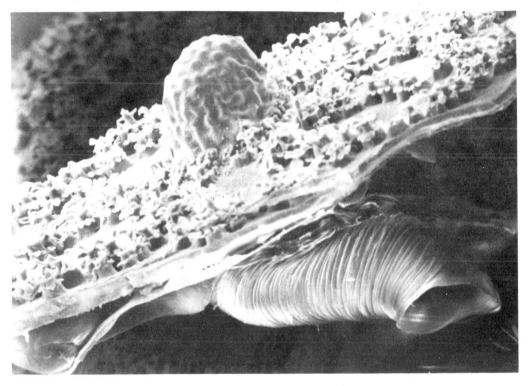

Scanning electron micrograph of the anterior spiracle and primary tracheal tube of a fly **puparium** (*Microdon piperi*, Diptera: Syrphidae), notice the circular reenforcement (taenidia) of the trachea. Magnified about 80 times.

38. Cannibalism

A few human societies have practiced cannibalism as a part of their culture. This was not ordinarily associated with obtaining nourishment but rather had religious or social significance. However, in a few instances, some human cannibalism has actually been for the purposes of nourishment and survival. One of the best known cases involved the Donner party trapped in a California mountain pass by an early snow fall in the 1800's.

Cannibalism by insects is fairly common. For example, lady beetle (Coccinellidae) larvae will consume adjacent eggs containing brothers and sisters. Horse fly (Diptera: Tabanidae) larvae also will voraciously attack and consume nearby siblings. Cannibalism allows these larvae to obtain energy until they can find other prey and reduces competition for available prey. This becomes particularly important when prey are relatively scarce (Dixon 1959). Newly hatched lacewing (Neuroptera: Chrysopidae) larvae also are voracious upon hatching, but to reduce cannibalism the eggs are always laid on the end of erect stalks (Richard & Davies 1977).

Queen and worker ants and bees will produce eggs that are laid for the sole purpose of furnishing nourishment to the brood or other colony members. These are called trophic eggs. Thus, the queen of a newly established fungus ant colony feeds her brood trophic eggs to maintain them until the fungus grows sufficiently to be consumed. Workers in stingless bee colonies also lay small trophic eggs that are consumed, mostly by developing larvae, but also by the queen (Wilson 1971). Yellowjacket wasps too frequently pull dead or unhealthy larvae from the comb, masticate them thoroughly, and then feed them to other larvae (Akre 1982).

Green lacewing (Neuroptera: Chrysopidae) eggs have long stalks to hold them safely above the leaf surface, the stalk didn't prevent this lacewing larva from cannibalizing an egg.

Temperature affects nearly everything we do. Because human enterprise and activity are so dependent on the weather we have television channels devoted exclusively to meteorology. The weather conditions reported on TV usually contains maximum and minimum temperatures for the day as recorded by very sophisticated thermometers.

Insects, being cold blooded, are even more sensitive to changes in temperature than people. Some can even be viewed as living thermometers. For example, the rate of chirping by house crickets (Orthoptera: Gryllidae) is related to temperature; it is fast paced during warm weather and slows down when the temperature drops. An enterprising naturalist devised a method of using the rate of chirps by a cricket to determine temperature. For the house cricket you can calculate the temperature with this equation:

$$T = 50 + \frac{N-4}{4}$$

T is the temperature in Fahrenheit and N is the number of chirps per minute (Frost 1959).

The entire life cycle of some insects is synchronized by temperature. For example the emergence of adult codling moths, a major pest on apples, in spring can be predicted by the accumulation of days (**degree-days**) when the temperature is above a known threshold temperature. Orchardists can use this information to time the application of pesticides to the period of peak adult codling moth activity.

Other insects can also be used as living thermometers. Human body lice will abandon a host when body temperature reaches 104° F. They will also abandon a corpse after it cools (Joklik & Willett 1976). Honey bees maintain the inside of their hives at a steady 85° F, but will increase it to 95° F to allow optimal brood development.

40. Full of It!

We use the slang expression, "You're full of it", when we perceive that a person is spouting misinformation. Translated into entomological jargon the phrase could perhaps be stated as, "You're talking like a larval bee." The connection between the two is that larval bees do not excrete their body wastes until they pupate, but instead hold these wastes in a blind gut called the **meconium**. The meconium is shed only at pupation (Borror et al. 1989).

41. Kamikaze

The Japanese term kamikaze, or divine-wind, first came into usage during World War II. In a desperate act to win the war Japanese pilots flew suicide missions by intentionally crashing their planes, laden with explosives, into the ships of Allied forces. These pilots considered it an honor to die for their emporer and regarded their lives as a small price to pay for the defense of their homeland.

Insects also have their kamikaze. Termites in the genus *Globitermes* have large mandibular gland reservoirs from which they can eject a liquid defense secretion that congeals on contact with air to trapping both the termite and the intruders. Sometimes the contractions are so violent that the abdomen bursts and the defensive fluid is splattered about (Wilson 1971). A similar defensive posture is assumed by some ants (*Camponotus* (*Colobopsis*)) in southeast Asia that burst as a defense against predators (Maschwitz & Maschwitz 1974).

77

42. Puffed up or inflation of the Body

A number of vertebrate animals can expand their bodies to temporarily become larger. A common example is the puffer fish that can greatly expand itself thus exposing numerous sharp spines which deter predators. Another well known example concerns lizards that run into crevices in rocks where they expand their bodies so would be predators cannot pull them out. Other animals such as birds and small mammals enlarge their apparent size by fluffing out their feathers or fur so that they appear larger to predators or rivals.

All these types of behavior were preceded by insect body expansions. In the most usual instance body expansion occurs during molting when the insect swallows air or water to become as large as possible so that later, after their cuticle hardens, it can grow into the new larger skin (Gillott 1980, Romoser 1981). While this case involves actual enlargement of the body, many examples occur among insects where they rapidly open their wings so they appear much larger to potential predators. Frequently this movement will reveal bright colors or eyespots that further startle the predator.

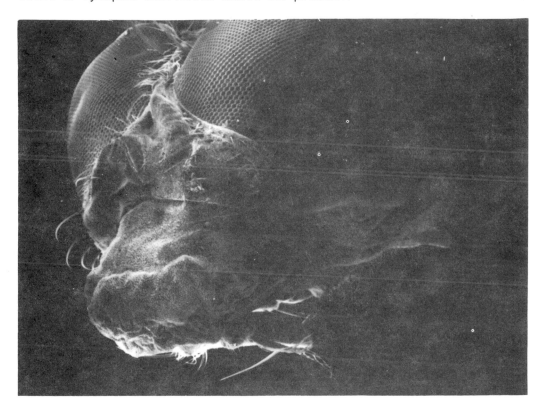

Scanning electron micrograph of the inflated **ptilinum** of a fly (Diptera) protruding from the front of its head. The fly uses the ptilinum to force its way out of the puparium.

A number of insects and one mite live on marijuana (*Cannabis sativa* L.) (Smith 1986) and other insects inhabit coca trees. Thus, insects made use of cocaine and marijuana long before humans discovered their pharmaceutical properties. In fact flesh fly (Diptera: Sarcophagidae) larvae develop faster on corpse tissue containing cocaine (Goff et al. 1989). This was somewhat unexpected but greatly influences the establishment of time of death by forensic entomologists assisting in the investigation of criminal cases involving drug related deaths.

44. Following Signs

Many years ago an old beekeeper from British Columbia said that the entire life of a bee is spent following signs. While it may be simplistic to refer to the behavior of bees in these terms, the statement has a basis in fact. One of the strongest signs or signals used to direct bees are "traffic signs displayed on flowers" more appropriately referred to as **honey guides**. Many flowers have ultraviolet reflecting patterns on their petals which guide bees to the center of the flower where the nectar is located. This is also an area where the bee is most likely to come into contact with the plants pollen which facilitates pollination. The honey guides direct the bees to search the pollen sites for nectar and thus force them to carry pollen from flower to flower (Von Frisch 1966, 1967).

81

In addition to simply being eaten by predatory insects, some insects have annoyance type problems with insects similar to problems that are experienced by people and other animals. For example, biting midges (Ceratopogonidae), of the genus *Forcipomyia* , act as blood sucking ectoparasites by imbedding their mouth parts into mantids (Mantodea), walkingsticks (Phasmida), dragonflies (Odonata), alderflies (Neuroptera: Sialidae), lacewings, beetles, moths, crane flies (Diptera: Tipulidae), and even mosquitoes (Diptera: Culicidae) (Borror et al. 1989).

Some insects and other arthropods will hitch a ride on other insects, this is called **phoresy**. A good example of a phoretic insect is a tiny parasitoid wasp in the Family Scelionidae. This wasp will ride on a female grasshopper and get off after the grasshopper lays eggs which the wasp then attacks. Other phoretic arthropods include bird lice (Phthiraptera) and pseudoscorpions (Pseudoscorpiones).

The torsalo, or human bot fly (*Dermatobia hominis*, Diptera: Oestridae), develops as a maggot within the skin of a wide diversity of mammal hosts. The adult female fly has a fascinating way of delivering her eggs to the intended host. She captures other flying insects, such as a mosquito, in the air, attaches a dozen or so eggs to it and then lets it go. Later when the mosquito lands on a warm host to bite, these eggs hatch and the tiny maggots bore into the skin.

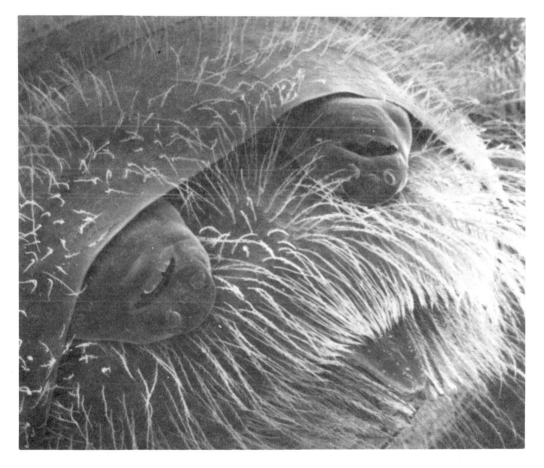

Scanning electron micrographs of twisted-wing parasites (Strepsiptera: Stylopidae) protruding from between the abdomenal segments of a bee.

46. Armor

Insects have an exoskeleton, this means that their skeleton or body framing is on the outside of the body. The hard exoskeleton gives most insects some degree of protection. However in many insects, particularly beetles, the exoskeleton is very hard, or **sclerotized**, to protect them from physical damage. However, insects have a still closer similarity to armor as used by knights in the middle ages. Caddisfly larvae (Trichoptera) build armor cases out of sand grains, pebbles, and vegetation for protection. Caterpillars, such as some carpet moths and bagworms (Lepidoptera: Psychidae), build similar cases (Borror et al. 1989). A good example is found in larvae of the snailcase bagworm, a serious agricultural pest, which produce extremely strong cases made of soil particles(Suomi 1988). Carrying this type of defense even farther are beetle larvae that use a shield made of their own feces to defend themselves from predators. They thrust this shield into the face of potential predators to discourage them when they attack (Eisner et al. 1967). Scale insects secrete a waxy material that protects them from weather and parasites. The secretions of *Laccifera lacca* (Kerr) (Homoptera: Kerridae), are gathered as stick lac, processed, and sold as shellac (Metcalf et al. 1962).

A heavily armored dung beetle (Coleoptera: Scarabaeidae) with pronounced tibial spines.

47.   Vampire

Tales about vampires have been popular ever since the author, Bram Stoker, created Count Dracula. Stoker's Count Dracula has evolved into the hollow tooth villian depicted in so many movies and other popular media. Tales of Dracula continue to be filmed even today. Vampire bats actually exist in Central and South America, but they are rather small flying mammals that cut animals with their tiny razor sharp teeth and then lap the blood. They are a nuisance in some areas of the tropics because they attack cattle and spread rabies.

The original vampires with "hollow teeth" for blood sucking were the insects. Such as lacewing larvae that suck the fluids of their insect victims through hollow jaws. Horse fly larvae can also be considered vampires as they have pores in the tips of their mandibles to inject toxins and enzymes into their prey (Philip 1931). Some blue bottle fly maggots (Diptera: Calliphoridae) will feed on the blood of nesting birds (Sabrosky et al. 1989), another, the Congo floor maggot (*Auchmeromyia luteola*, Diptera: Calliphoridae), sucks the blood of people in Africa who sleep on the floor of their huts (Oldroyd 1964).

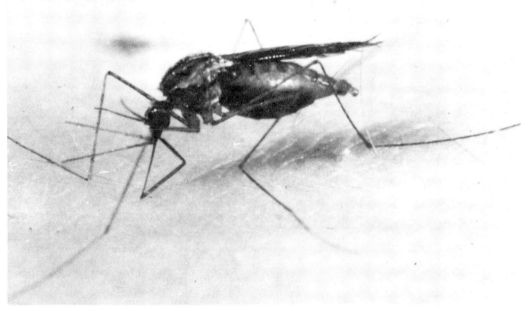

A partially engorged mosquito (Diptera: Culicidae) feeding on a mans arm.

The pollution of our lakes, rivers, and streams with noxious chemicals is of national and international concern. However, on a much smaller scale, some of the original polluters of water are aquatic insects. All insects must excrete nitrogen waste accumulated as products of protein metabolism. Most terrestrial insects convert these waste products into an insoluble and therefore non-toxic compound, uric acid, before they are excreted. Aquatic insects however, discharge ammonia, a highly toxic material, directly into water. Since these insects live in an aquatic habitat, they are constantly awash in a cleansing solvent (water) which quickly dilutes and dissipates this toxic ammonia waste. Converting metabolic wastes into uric acid is an expensive process in terms of energy, so aquatic insects have an advantage over their terrestrial counterparts (Romoser 1981).

A water boatman (Hemiptera: Corixidae) resting in *Elodea*.

49.   Hypodermic

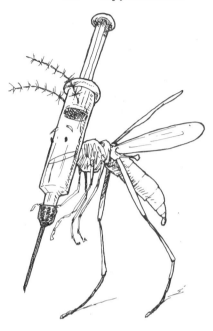

The use of hypodermic syringe and needles for the administration of medicines or for drawing blood a common practice, as is, unfortunately, the use of the hypodermic syringe to "shoot up" illegal drugs. The syringe is armed with a needle, a sharply pointed fine caliber tube, used to pierce the skin. The needle is used with a syringe which works on the principle of differential pressures in the blood vessels and the syringe. The original hypodermic belongs to the insects. The mouthparts of most blood sucking and plant sucking insects are sharply pointed elongated structure that interlock to form a piercing tube. Food is carried up the mouthparts as the result of suction created by the expansion of diaphragm-like pumps in the head of the insect. Such hypodermics can be found in many bloodsucking insects such as bed bugs (Hemiptera: Cimicidae), mosquitoes (Diptera: Culicidae), blackflies (Diptera: Simuliidae), and stable flies (Diptera: Muscidae). All inject materials to inhibit blood clotting as they feed (Harwood & James 1979). Similar hypodermic-like mouthparts occur among aphids (Homoptera: Aphididae), scale insects (Homoptera: Coccidae) and most true bugs (Hemiptera).

90

50. Acrobats

Traditionally circus shows feature acrobats demonstrating their remarkable skills at various forms of tumbling or balance. Again, the insects were first. Flies (Diptera) of the family Piophilidae, occasional pests in dried meat and cheese, are called cheese skippers because the larvae can catapult themselves through the air, sometimes up to several inches. They accomplish this by bending their body nearly double, grasping their rear end, tensing their muscles and then suddenly letting go. As their body straightens and strikes the substrate they are thrown high into the air. This is probably an adaptation to escape predation (Borror et al. 1989, Metcalf et al. 1962).

Mexican jumping beans, are seeds of *Sabestiania* plants infested with the caterpillar of *Cydia deshaisiana* (Lucas) (Lepidoptera: Tortricidae). The caterpillar which is enclosed in the bean will thrash about when the seeds are warmed in a human hand or placed in the sunlight causing the beans to tumble (Borror et al. 1989).

91

51. Blinking Neon

Flashing neon lights are a sure fire attention getter that have been used in advertisements for years. Perhaps the extent of these garish displays was first brought to public attention during the gasoline crisis of 1976 when lights of the gambling casinos of Las Vegas and Reno were threatened with shutdown by the power companies as a means of saving oil and countless megawatts of electricity.

Long before people used flashing lights as attention getters, fireflies (Coleoptera: Lampyridae) of temperate regions employed a light produced in the posterior segments of their abdomens to attract a mate. Various species even developed a morse code of specific signals that attracted only others of their species (Carlson & Copeland 1978, 1985, Lloyd 1971, 1979). Female fireflies of the genus *Photuris* went one step further by mimicing the signals of another species to lure males to be eaten (Lloyd 1965). An additional twist to this firefly story was discovered in Asia in the form of congregational trees 35 to 40 feet tall containing great masses of fireflies that synchronize their flashes (Buck & Buck 1976) (see Chapter 76). This synchronized flashing presumably enhances mating.

The chemical reaction responsible for producing light in fireflies is an enzyme (luciferin/luciferase) based system powered by **ATP** (Adenosine Tri-Phosphate), an essential molecule in the physiology of earth lifeforms. This reation is being used by NASA as a means of determining the possible presence of life on other planets. A mixture containing luciferin-luciferase is included in a container aboard exploratory spacecraft. After the spacecraft lands a mechanical arm can then be used to scoop some of the substrate (soil) into this container. If ATP is present, the system will be activated, sending a signal back to earth that life, at least as we know it on earth, may exist on the planet.

52. Velcro

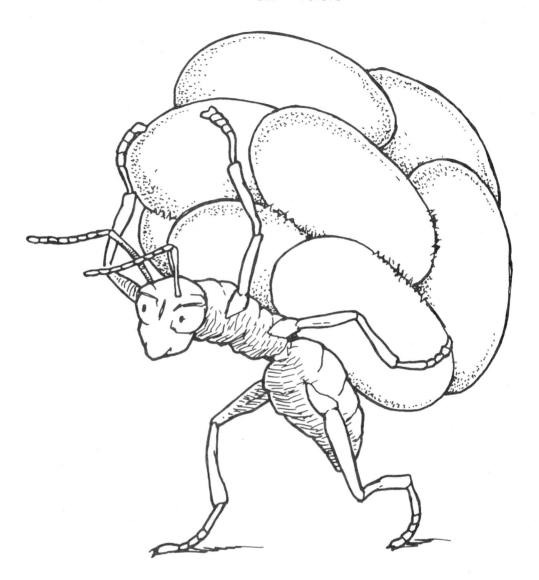

Velcro was developed in 1948 by a Swiss engineer, George deMestral, who was curious why cockleburs stuck to his socks and his dog after a walk in the woods. He examined the burs with the aid of a microscope and discovered they were covered with hundreds of tiny hooks which attached themselves to anything loopy. He invented a method of duplicating the hook and loop configuration in nylon and named the product Velcro. The basic patent expired in 1978, and dozens of manufacturers throughout the world currently produce variations of this device.

Not only plants but also insects developed the same type of morphology of hair and spines. Some ant larvae are very hairy causing them to cling together allowing workers to pick them up in groups rather than singly (Wheeler 1910). Workers can then transport bundles of larvae to different places in the nest. Under high magnifications these hairs look remarkably like the hook portion of velcro.

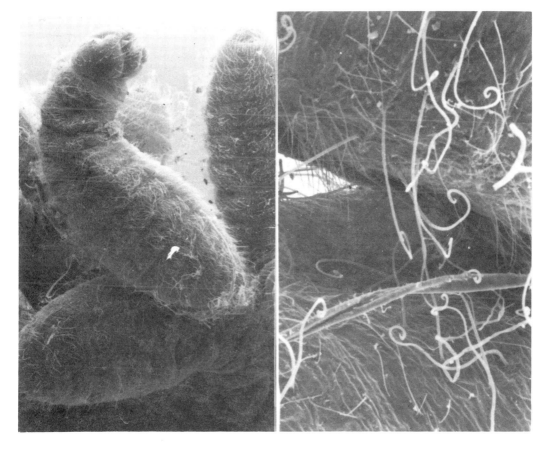

Scanning electron micrographs of ant larvae (magnified about 40 times) and a close up of the velco-like hairs (setae) that cause them to cling together (magnified about 250 times).

53. Superball

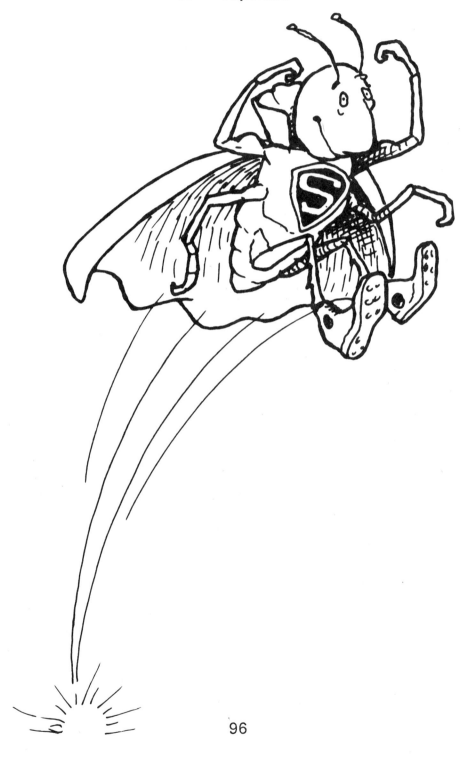

When chemists found a material that could store mechanical energy of compression efficiently, it was manufactured into small "rubber" balls for children. These "super balls" were extremely resilient and when bounced, rebounded with much greater vigor than ordinary rubber balls. However, a similar material has been used by insects for eons (Anderson & Weis-Fogh 1964, Ross et al. 1982). Have you ever wondered how tiny fleas can jump so high? Insect physiologists studying the leap of the flea (Siphonaptera) discovered that these insects have a super elastic protein called **resilin** in the thorax above the hind leg. Resilin is able to store 97% of the energy of compression. The muscles of the thorax and the legs compress this protein so it rebounds with great energy when released catapulting the flea to great heights in its leap (Rothschild et al. 1973, Duplaix 1988). Storage of kinetic energy in resilin is also very important in the functioning of the **"click mechanism"** associated with insect flight when muscle tension is released into the power stroke of the wings (Kerkut & Gilbert 1985) (see Chapter 8). Resilin is an important component in the highly efficient use of energy by fast flying insects.

The largest flea in the world, *Dolichopsyllus* sp. (Siphonaptera: Ceratophyllidae), an ectoparasite of beavers.

54. Original Big or Swelled Heads

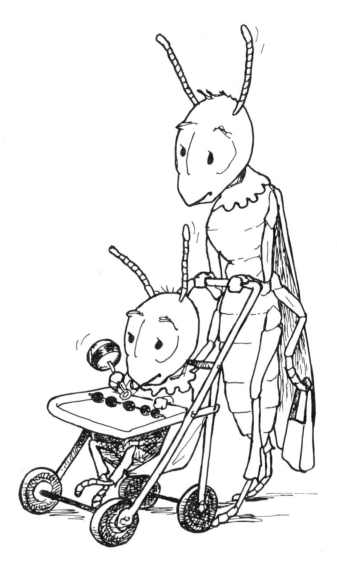

In American slang the expression to have a "swelled head" denotes that the individual in question has an inflated value of his or her worth. The head of a young child too is much larger proportionally than that of an adult. Insects actually have big heads too, nowhere is this more evident than in newly hatched grasshoppers (Orthoptera: Acrididae) nymphs. Their heads are so large that they are out of proportion to the rest of their body. However, this is a case of allometry, or differential growth of certain body parts. As the grasshopper passes through successive molts on the way to becoming an adult, the head ultimately assumes the correct proportion to the body (Richards & Davies 1977, Elzinga 1987, Matheson 1957). The same as with people.

A family of flies (Diptera), the Pipunculidae, have very large heads, comprised mostly of the compound eyes, in relation to their bodies, and thus their common name of big-headed flies (Borror et al. 1989). Many other flies posses a structure called a **ptilinum** which aids the adult during emergence. This structure is an eversible sac located internally at the front of the head. When emerging from the pupal case, the ptilinum is repeatedly expanded and contracted by forcefully being filled with body fluid, allowing the insect to push its way to freedom. When the ptilinum is fully everted the insects head is swollen to twice its normal size. After the insect is free the ptilinum is withdrawn inside the head and never seen or used again.

A big-headed fly (Diptera: Pipunculidae), the head is almost entirely composed of large compound eyes.

Polyesters are in great demand as fabrics for clothing, but again, insects did it first (Hefetz et al. 1979). Bees in the genus *Colletes* (Hymenoptera: Colletidae) make brood cells in the soil to raise their young. The female bee lines these cells with a waterproof, transparent membrane. This material, produced by a gland (**Dufour's**) in the abdomen of the bee, is reported to be a naturally occurring linear polyester. Additional studies suggested that salivary enzymes, applied from the mouth of the female bee, are necessary to create the polyester chains from lactones in the Dufour's gland (Torchio et al. 1988). If this is not fantastic enough, it was further reported that a parasitic bee preys on *Colletes* larvae in the brood cells. When these parasites slit the polyester-lined cells to lay their own eggs, a material is deposited along with the egg that dissolves the polyester lining. However, it quickly rehardens to protect the newly laid egg (Torchio & Burdick 1988). Thus, insects not only manufacture polyesters, but one even produces a secretion that acts as a repair kit for the cut it has made in the brood cell lining (Sherman 1989).

All human societies have beggars of one sort or another on street corners beseeching others for food or money. It is no different with social insects. Because social insects usually have stores of food, other insects constantly try to exploit their larder. Perhaps the closest analogies to beggars are found among the myrmecophiles or ant lovers, insects that live in colonies of ants. One example concerns small crickets of the genus *Myrmecophila* (Orthoptera: Gryllidae) that live with thatching ants. These crickets are able to duplicate the antennal touch pattern that worker ants use to beg food from one another. The crickets will cautiously approach a worker ant, tap the ant with its antennae in a sequence that ordinarily causes the workers to feed the beggar. However, sometimes the sequence is performed incorrectly, and the cricket has to be very swift to avoid being attacked and killed. Not all are successful (Henderson & Akre 1986a). Other examples of **inquilines** (nest dwellers) that beg for food include silverfish (Thysanura), cockroaches (Blattaria), and beetles (Kistner 1982, Wilson 1971).

A small cricket (*Myrmecophila*) found with several species of ants (*Formica* and *Camponotus*). These crickets have broken the code used by the ants for communication so they can successfully beg for food.

Tool use involves the manipulation of an inanimate object, not internally manufactured, to improve the efficiency of the animal in moving some other object (Alcock 1972). Another definition states that a tool is an object separate from an animal's own body that is used to extend the animal's capabilities (McMahan 1983).

Tool use by social insects was first reported by Lin (1964) for the pavement ant (*Tetramorium* sp., Hymenoptera: Formicidae). Worker ants use soil as a weapon to attack bees nesting in the soil near the ant's nest. These ants drop soil particles down into the entrance of the bee nest to induce the bee to come

104

to the soil surface. Of course, once the bee is out more worker ants attack and kill it. Since then similar behavior has been reported a number of times (Fellers & Fellers 1976, Fowler 1982, McDonald 1984, Moglich & Alpert 1979, Schultz 1982). Another type of tool use by ants involves the manipulation of soil or detritus particles to soak up honey for return to the nest or to move the honey elsewhere in the nest (Barber et al. 1989).

Among nonsocial insects only a few instances of tool use have been recorded. In one case, a bug (Hemiptera) in Costa Rica uses a dead termite to attract additional prey. The first termite must be captured by stealth, but then the bug waves the dead termite about to attract another termite, and it, in turn, is pierced by the mouth parts of the bug and sucked dry. This continues, termite-after-termite, until the bug is satiated (McMahan 1983).

One species of wasp (Hymenoptera: Sphecidae) uses a pebble as a tool to pack sand over its nest. The female wasp catches prey which is placed into a burrow she has excavated. She then lays an egg on the prey and refills the hole using a pebble, held in her mandibles to pack the soil (Brockman 1985). The egg hatches, the larva consumes the prey, pupates, and ultimately digs itself out as an adult wasp. The use of the pebble to pack the soil by the insect is analagous to the use of mechanical compactors which we use to smooth and compact soil at construction sites.

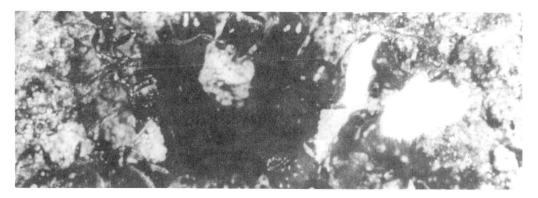

A worker of the pavement ant dropping a soil particle on an alkaili bee to induce the bee to leave the burrow so it can be killed.

105

Primates are somewhat unique with regard to other animals because they have a thumb which opposes the fingers and can be used for grasping and manipulation of objects. The ultimate use of the opposable thumb, of course, comes only when combined with intelligence, therefore the most refined usage is by the human primate.

Lice (Phthiraptera) can also be considered to have an opposable thumb if a little "poetic license" is permitted. The tarsi of sucking lice are one segmented and have a single large claw which fits against a thumb-like protrusion at the end of their lower leg (**tibia**). This opposable thumb and claw is used to cling to the hairs of the host (Borror et al. 1989)

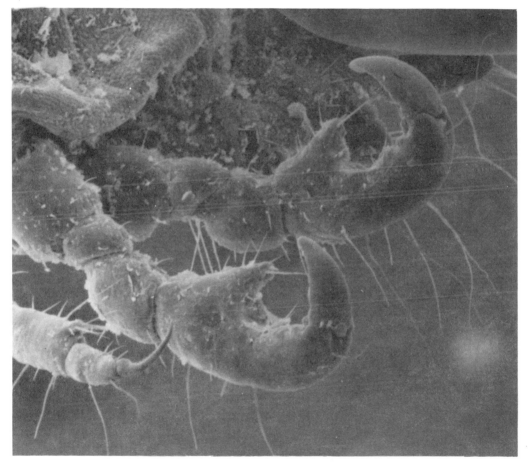

The claws of a crab louse (*Pthirus pubis*(L.), Phthiraptera: Pthiridae) magnified about 150 times.

Mimicry in its simplest form is appearing or behaving like another animal. Thus children sometimes mimic their parents in dress and behavior, and talking birds mimic some of the sounds that they hear. However, the absolute masters of mimicry and deception are the insects.

Mimicry abounds in the insect world (Rettenmeyer 1970, Pasteur 1982). There are two more common forms of mimicry. **Batesian**, in which an palatable insect mimics a model that is distasteful or has a sting, and **Mullerian**, in which a number of distasteful or stinging insects look alike for mutual protection. Undoubtedly one of the most famous cases of Batesian mimicry involves the edible viceroy butterfly (Nymphalidae) which mimics the inedible monarch butterfly (Danaidae). Predators learn quickly that monarchs will cause them to vomit if eaten. Due to its monarch-like appearance the viceroy is also protected from predation(Brower 1969, Erlcih & Raven 1967). Unfortunately, new information suggests the viceroy is palatable and therefore is not a Batesian mimic but rather a Mullerian mimic(Ritland & Brower 1991). However many other examples of Batesian mimicry do exist in the insect world. Many harmless insects, such as katydids, moths, and flies, are mimics of bees and wasps.

Mullerian mimicry is also relatively common. A good example involves the tiger stripe butterflies of the Neotropics. The group is a complex of many species from three families of butterflies, the butterflies are all very close in appearance and are all unpalatable (Papgeorgis 1975).

A more unusual type of mimicry is that exhibited by ant guests, particularly beetles, that look like their ant hosts. This is called **Wasmanian mimicry** (Rettenmeyer 1970). In some **myrmecophilous** flies (Diptera: Syrphidae) mimicry of an ant host is even extended to mimicing the integumentary chemicals important for recognition of colony members (Howard et al. 1990).

60. Bungee Jumpers

Some people claim bungee jumping is one of the greatest experiences of
your life. Tie an elastic cord around your legs and jump from a tall bridge! The
cord will safely stop your descent, and the effect is one of great exhilaration.
However, the original bungee jumpers are, of course, insects and spiders.
Caterpillars of the apple-and-thorn skeletonizer (Lepidoptera: Choreutidae),
Choreutis pariana (Clerck), frequently dive out of trees when they are
threatened by predators only to be saved by their "bungee cord" of strong silk
spun from their salivary glands (Metcalf et al. 1962). Black fly larvae living in
fast flowing streams also spin a silken safety cord or bungee to stop them from
being swept away when they switch positions in a stream. They simply play out
the line until they can re-fasten to a rock with their posterior prolegs (Oldroyd
1964). Many spiders, of course, employ similar types of escape behavior.

110

61. Robbers

Insects are experts at stealing from other insects. There is a family of flies
(Diptera), the Asilidae, commonly called robber flies, which were thought to be
experts at divesting other predators of their hard earned prey (Borror et al.
1989). Despite the name however, most species of robber flies are simply good
predators, not thieves. There are certain blow flies (Diptera: Calliphoridae,
genus *Bengalia*) that are robbers. They wait beside raiding columns of army ants
or carpenter ants and simply steal the prey away from the worker ants returning
to the nest. Since the ants are flightless, the flies accomplish their robberies
with relative impunity and then fly away to eat at their leisure (Hölldobler &
Wilson 1990). Triatomine (Hemiptera: Reduviidae) bugs will steal blood
(**haematoclepty**) from other bugs swollen with a blood meal by inserting their
mouthparts into the other insects distented gut. There is no apparent resistance
or harmful effect on the donor (Miller 1971).

111

Humans have developed some extremely sophisticated time measuring devices that are far removed from Stonehenge, sundials, and hourglasses of yesteryear. Most adults in the United States now wear a quartz watch. Even relatively inexpensive models are accurate to within several minutes a month. Without clocks and watches the biorhythms of humans operate on a cycle that is close to the 24 hour daily cycle (**circadian** = about 1 day). Even these rhythms are lost slowly when not reinforced by observing a normal day-night cycle. People in a cave environment, subjected to constant light, ultimately lose their natural circadian rhythms of sleeping and eating.

Insects also make use of internal clocks for controlling many of their rhythms, from mating to overwintering (**diapause**) in order to survive adverse conditions (Saunders 1976, 1982). Honey bees have accurate clocks that enable them to detect differences in the position of the sun to within a few minutes. However, bees do have trouble orienting at the equator at noon, with the sun directly overhead, but for only a few minutes (Von Frisch 1967) (see Chapter 18). Ants also have sophisticated navigation systems that use very accurate biological clocks (Holldobler & Wilson 1990).

63. Baskets and Pots:

One of the worlds oldest form of basket are the pollen baskets or **corbiculae** found on the hind legs of honey bees. Bee workers pack these baskets full of pollen before returning to the hive where they unload their burden of honey and pollen. Other insects make pots of various shapes. Potter wasps (Hymenoptera: Eumenidae) store their prey in clay pots fashioned by the wasp using it's mouthparts and mud. Then the female wasp seals the pot after laying an egg inside the pot with the paralyzed insect prey. The hatching larvae then feed on the prey and complete development inside the sealed pot. Finally they chew away the seal and emerge as adult potter wasps. Other pot makers include certain species of burying beetles (Borror et al. 1989).

64. Ballooning

Some of the first flights of humans were in baskets suspended below air tight bags of silk or other fabric that were filled with lighter than air or heated gases.

Insects are also balloonists. Smaller larval forms, particularly caterpillars of moths, spin threads or parachutes of silk that catch the wind and transport the insects for miles. For example, the newly hatched larvae of gypsy moths (Lepidoptera: Lymantridae) drop from the tree tops on strands of silk to be caught and transported by the wind. In addition to silk, the lateral hairs (**setae**) on the bodies of the caterpillars also help their buoyancy (Atkins 1980, Frost 1959). Spiders also use balloons for dispersal and in the spring hundreds of small newly hatched spiderlings are transported for great distances by this means (Borror et al. 1989).

People pride themselves on advances in preserving and storing food. Some of the techniques in current use are freeze drying and irradiation for sterilization of food. Insects also have an effective method for food storage. Solitary wasps sting their prey in the thoracic ganglia of the nerve cord so that they are totally paralyzed but are still alive. The prey, for example a spider, is then eaten by a developing wasp larva. The prey does not decay, but is in a zombie-like, living dead state until it is consumed (Steiner 1983).

116

66. Head Screwed on Straight?

Perhaps at no other time is it more appropriate to bring up the slang term "is your head screwed on straight." This question may infer that the individual in question is simply somewhat confused or perhaps under the influence of mind-altering drugs or alcohol.

However, in regard to insects it is important for them to be able to sense exactly where their head is in relation to the rest of their body. This is particularly true with dragonflies. Their ability to sense the position of their head in relation to their feet allow them to catch their prey in a basket formed by their legs. As the dragonfly swoops in to make the catch it is essential that the head and the basket are correctly aligned. This is possible due to the presence of two pads covered with sensory hairs located on the dragonfly's neck (Carthy 1958). These pads allow the insect to accurately align the head to the body. Similarly, honey bees also have sensory pads on the neck (actually the **prothorax**) to align the head and body for orientation of the waggle dance to the correct angle of the sun as the angle is converted to a response to gravity in the darkened hive (Von Frisch 1967, Winston 1987).

67. Communes

Semi-social living groups among humans known as communes arose during the protest movements of the 1960's. Relationships in the commune tended to be very elastic and non-permanent among the commune inhabitants.

Some insects have a similar type of communal organization. These insects group together on a common nesting ground. Communal nesting, is a loose association of individuals with very little sharing of resources. Alkali bees are one example of insects with such an association. However, members sometimes become confused to the point that they provision nests belonging to others (Wilson 1971).

68. Alphabet

The alphabet as used by humans to spell words and sounds was illustrated by insects even before the idea of an alphabet was conceived. Actually, the letters of the entire alphabet, A-Z, as well as the numbers 0-9, occur on the scale covered wings of different butterflies, within the insect order, Lepidoptera (Lipske & Sandved 1988).

119

69. Humming

The death's head moth (Lepidoptera: Sphingidae), featured in the movie "Silence of the Lambs", produces sound by means of an airstream drawn in through its tube-like **proboscis**. The in-rushing air causes the flap-like **epipharnyx** to vibrate producing a pulsed airstream and a low pitched sound. When the air is expelled a high pitched whistle is produced (in slang terms, blowing it out their nose!) (Atkins 1980, Haskell 1964). These sounds are produced when the moths are handled and may act as a warning to potential predators. The same apparatus is reputed to produce sounds which mimic the piping sounds of a virgin honey bee queen allowing the death's head to rob bee hives of honey. However, this ruse is not without risk, many dead moths are found inside the hives where they have been killed by honey bee workers (Wynne-Edwards 1962).

70. Fishing Net

The development of fishing nets allowed fishermen to catch greater numbers of fish with less effort. When nets were incorporated into traps, fish could be caught in the absence of the fisherman allowing them to invest more time into other activities. Fishing nets are still used extensively today. They range from small hand nets thrown from the shore to the controversial drift nets, miles long, floating in the high seas.

Insects also use fishing nets. Many species of caddisflies (Trichoptera) live attached to the bottom of rocks or the stream bed in silken tubes or case-like retreats constructed from pebbles and sand grains. These insects construct nets which catch food from the water as it flows through the net. Some nets are trumpet shaped while others are more sheet-like (Borror et al. 1989). The mesh of the net is very uniform and under high magnification looks very much like many nets of human manufacture.

121

Most insect colors, such as those found in the wings of butterflies or the eyes of horse flies, are the result of structural characteristics that selectively refract and reflect light into gorgeous colors. However, some colors are due to chemicals manufactured by the insects. The cochineal insect (Homoptera: Dactylopiidae) are red due to the presence of cochineal or carminic acid (Nauman 1991). Cochineal is a beautiful carmine-red dye that is obtained from the bodies of the cochineal insect, *Dactylopius coccus* (Costa), that occurs in the American tropics. This dye and similar dyes are used by insects, probably as warning coloration and as a defensive material if attacked by predators. These dyes are also used by humans. Cochineal dye was processed as a powder and was once a major export item from Spanish dominated Central America. It was used by the Aztecs of Mexico long before the Europeans arrived, and is still in limited use today. Today aniline and other synthetic dyes are so common that most people have forgotten that insects were once an extremely important source of dyes for both foodstuffs and fabrics. Dyes such as Turkey red are extracted from insect produced galls, and other dyes obtained from galls are used in dyeing wool and skins or for making permanent inks (Metcalf et al. 1962).

72. Break Sound Barrier

Colonel Chuck Yeager was the first human with "the right stuff" to fly an airplane faster than the speed of sound (750 m.p.h.). Today, breaking the sound barrier is "old hat" and even some commercial airliners are capable of reaching speeds that were once considered the realm of only experimental aircraft.

Insects were reputed to be the first to break the sound barrier, at least if you believe everything you read. C. H. T. Townsend, a famous scientist who worked on flies, estimated that male deer bot flies (Diptera: Oestridae) were capable of speeds of up to 800 miles per hour (Cole 1969)! Exceeding the speed of sound by a considerable amount. Of course, this statement ignores the fact that no insect could physically withstand these speeds. Their wings would break off long before this velocity was attained, and the insect would literally explode into pieces. Thus, while insects have been credited with this feat, it simply is not true. Male deer bot flies probably do not exceed speeds of 50-70 m.p.h.

73. Cannon

The water cannon is currently a well known riot control device. The cannon shoots a spray of water, onto a crowd under pressure great enough to simply knock people over. This, combined with wetting by the water, tends to dampen the ardor of rioters.

The original canon is possessed by bombardier beetles, *Brachinus* (Carabidae), that produce hot (100^0C) sprays of quinones when disturbed or attacked (see Chapter 21). These quinones are so reactive that they exist in the beetles only as precursers that are mixed only immediately prior to use. The quinones are sprayed from the beetles anus and can be aimed in different directions. Upon contact with the air the quinones produce a sound, and a visible cloud of chemical vapor. The sequence reminds one of the firing of a black powder cannon.

After two world wars, multiple police actions, and limited military encounters, the people of the world are still engaged in a deadly arms race which involves the manufacture and storage of a superabundance of all types of lethal weapons. While this behavior may seem to be uniquely human, it is not. The original arms race probably started shortly after life began.

At first, life forms were soft bodied, but 500 to 600 million years ago in the Cambrian period of geological history, a relatively sudden transition took place during which many animals became heavily armored and spined (Stolzenburg 1990). Although no insects are known to have occurred during this time, these insect ancestors entered the "arms race" with a vengeance, and that race continues today. Many insects are heavily sclerotized (beetles) and also have the capability of delivering nasty bites (beetles, grasshoppers) or stings (wasps, bees, ants) to protect themselves from their enemies.

Another arms race occurs between insect plant feeders (**herbivores**) and their host plants. Plants with the more complex defense systems also have the greater diversity among their herbivores (Cowen 1990). The ecological term for these adaptations among insects and the counter adaptations among their predators, parasites, and food plants is called **coevolution**. An interesting coevolutionary relationship exists between certain species of plants and ants. The plants, called **myrmecophytes**, provide nesting sites for ants, in return the ants protect the plants from herbivores. Some of these plants have evolved an array of highly specialized structures for harboring ant colonies: for example *Myrmecodia*, **epiphytic** plants found in Southeast Asia, will form large hollow pseudobulbs which are occupied by *Iridomyrmex* ants (Wilson 1971). Perhaps the best known example of ant-plant **mutualism** are the bull's-horn acacia's of the American tropics which house *Pseudomyrmex* ants inside hollow thorns (Janzen 1966).

127

75.   Bridges

The problem of crossing rivers or chasms was solved by early humans with the construction of bridges from logs or vines. The art of bridge building progressed steadily from these times as both an art form and an engineering challenge (Bonanos 1992, Petroski 1992).

However, among the earliest bridge builders were the ants, particularly the army ants of the Neotropical regions. For example, *Eciton burchelli* (Westwood) forms bridges between vines and roots by using their own bodies. The ants simply hold onto each other to form a natural bridge that can then be used by the remaining members of the colony to speed their journey. Workers also build living flanges of more ants on vertical and horizontal surfaces of the bridge to widen their pathway (Rettenmeyer 1963).

76. Sexual Bias

In this day of affirmative action and equal treatment for both sexes in the work place it might come as somewhat of a shock to find out that insects are hopelessly biased toward females regarding sex. In fact, in some species of insects there are no males. Females reproduce **parthenogenetically** (see Chapter 77), producing viable eggs without being fertilized. Many bees, wasps, and ants (Hymenoptera) can control the sex of their offspring at will. They do this by controlling the release of sperm from the receptacle in which it is stored (spermatheca) (Gerber & Klostermeyer 1970). Fertilized eggs produce females, while unfertilized eggs always develop into males. This is called **haplo/diploid reproduction**. Because of this bias virtually all honey bees, social wasps, and ants are females.

129

The phrase "red light district" may have biblical origins. In Joshua, Chapter 2, the harlot, Rahab, helped two spies escape their pursuers by climbing out her window on a rope. When the spies returned with the conquering army, Rahab hung a scarlet cord from the window by which she helped the spies escape, as a signal for the invading army to spare her and her family. Since that time not only were red lights used in the window to signify houses of prostitution, but in New Orleans red brick dust was also spread on the sidewalks in front of these houses.

Although many insects can detect colors, most are blind to the red end of the visible spectrum. Therefore many studies of nocturnal insects can be undertaken using red lights without disturbing the insects. However, some fireflies (Lampyridae) actually use flashing red light as mating signals (McElroy 1964). Most other fireflies emit light that varies from green through orange (Lall et al. 1980) (see Chapter 51). This color, of course, can vary with temperature and other environmental conditions. In addition, some click beetles (Coleoptera: Elateridae) have abdominal lights that glow red (Lloyd 1971).

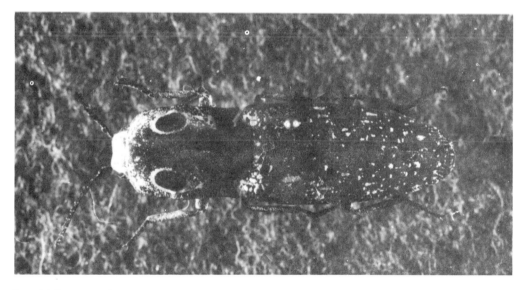

Eyed elaterid (Coleoptera: Elateridae) has pronotal and abdominal lights similar to those of fireflies. Some species have red lights.

It is, hopefully, unnecessary to go into any detail about the sexual antics of people. Suffice it to say that a number of types of human sexual behavior can definitely be called unusual or even "kinky." However, even here the human race has been "firsted" by a bunch of bugs. Bed bugs (Hemiptera: Cimicidae) use their sclerotized (chemically hardened) male organ (**aedeagus**=anterior extension of the phallus) to penetrate a female directly through her body wall for the deposition of sperm into specialized tissues (**spermalege**). Male bed bugs of the genus *Hotea* are equipped with bizarre large, heavily sclerotized genitalia which are half the size of the female abdomen. This "can opener" is used to tear their way through the vagina to deposit sperm directly into a sperm receptacle (**spermatheca**). Many male bed bugs have copulatory scars caused by the aedeagus of other males. If the "raped" male is *in copula* with a female as it is raped, sperm of the second male may find its way into the female via the first males sperm ducts and aedeagus (Thornhill & Alcock 1983) (see Chapter 81).

Many of the unusual adaptations are intended to secure for males the sole reproductive rights of a female. For example, sperm from a male dung fly placed into a female is pushed out by the next male and in some species of heliconiid butterflies the male chemically marks a female during mating to deter other males from mating with that female. Male walkingsticks take this strategy one step further and actually ride the female for days or even weeks as a living "chastity belt". At another extreme are the species of butterfly in which the male glues his genitalia to those of a female and the ceratopogonid midges (a tiny fly) in which the male leaves his severed copulatory organ inside the female to prevent other males from mating with the female. Other insects simply insert plugs into the female during the act of mating so other males cannot subsequently copulate with them.

Other cases of unusual sex among the insects abound. One **parthenogenic** (female able to reproduce without being fertilized) species of spider beetle even mates with males of another species in order to use the ejaculate as a nutritional boost. Females of a species of beetle found in the African desert may be overpowered and forced to mate with large numbers of males (National Geographic Film Special: The Namib). Then of course, there are the insects such as coccids that are self fertilizing **hermaphrodites** (Lloyd 1979, Engelmann 1970, Thornhill & Alcock 1983). The list can be extended almost indefinitely with occurrences such as the pseudocopulation of bees or flies with flowers, usually orchids (van der Pijl and Dodson 1966), and the attempted matings of mites with sand grains or nearly any other object of a suitable size.

133

79. Sexual Bondage

Even sexual bondage is a first when spiders (Arachnida) are included in the discussion. Males of *Pisaurina mira* (Walckenaer), before attempting to mate, tie the female's first and second pairs of legs with silk, and then the third and fourth and fourth pairs are held tightly in an embrace. The function of this somewhat bizarre behavior is presumably to reduce the chances of the male being attacked and killed by the females (Bruce & Carico 1988).

A similar type of bondage occurs in the behavior of mating mites (Acari). Males of two spotted spider mites guard quiescent immature females (**deutonymphs**) and web them over extensively. The male then stands over the virgin until she emerges as an adult, and they quickly mate (Penman & Cone 1974).

Insects also exhibit mating behavior similar to bondage. In most cases the male simply grasps a female by some portion of her anatomy and remains attached, sometimes for days, until the female is receptive to mating. Even then, in many cases, the male remains in position to prevent the female from mating with other males (Thornhill & Alcock 1983).

Machilis germanica (Verhoeff) (Microcoryphia: Machilidae), a primitive, wingless insect, uses a thread spun from the end of his abdomen to guide and push the female onto his sperm packets (**spermatophores**) that are secreted onto this thread (Englemann 1970).

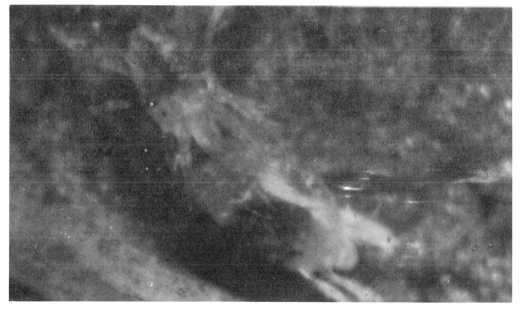

Male spider mite guarding a female deutonymph so he can mate with her as soon as she emerges. The male has even spun a web over her and will defend her against other males.

Through the wonder of modern medicine it is possible for humans to change sex, but only superficially. People that have undergone sex changes cannot perform the reproductive role of their new sex. Again insects did it first, and, in this case, better. A fly that lives with fungus culturing termites first passes through a male stage during which spermatozoa are formed, and then later transforms into a female stage and eggs (**oocytes**) are developed (Royer 1975).

Male bed bugs (Genus *Xylocoris*) are sometimes mounted by other males that inject them with sperm (Lloyd 1979) (see Chapter 77).   The sperm eventually finds its way into the sperm ducts of the mounted male and into the female with which the raping male did not copulate.   Codling moths that become overstimulated by the female mating pheromone may attempt to mate with other males, sometimes so vigorously that the heads or abdomens of the "mated" male are crushed by the claspers of the overstimulated male.

137

ORDERS OF INSECTS

There are 28 Orders of insects. Each order is sub-divided into Families, Family names can be easily identified because they always end in the suffix, -idae. The Families of insects mentioned in this book are listed alphabetically under the Order in which they belong.

Microcoryphia - Jumping Bristletails

    F. Machilidae

Thysanura - Silverfish

Ephemeroptera - Mayflies

Odonata - Dragonflies and Damselflies

Grylloblattaria - Rock Crawlers

Phasmida - Walkingsticks and Leaf Insects

    F. Phasmatidae

Orthoptera - Grasshoppers, Crickets, and

    Katydids
        F. Acrididae
        F. Gryllidae
        F. Tettigoniidae

Mantodea - Mantids

Blattaria - Cockroaches

Isoptera - Termites

    F. Macrotermitidae

Dermaptera - Earwigs

Embiidina - Web-spinners

Plecoptera - Stoneflies

Zoraptera - Zorapterans

Psocoptera - Psocids

Phthiraptera - Lice
    F. Pediculidae
    F. Pthiridae

Hemiptera - Bugs
    F. Cimicidae
    F. Corixidae

    F. Lygaeidae
    F. Reduviidae
    F. Scutelleridae

Homoptera - Cicadas, Hoppers, Psyllids,

    Whiteflies, Aphids, and Scale

    Insects
    F. Aphididae
    F. Cercopidae
    F. Chermidae
    F. Coccidae
    F. Dactylopidae
    F. Fulgoridae
    F. Jassidae
    F. Kerridae
    F. Psyllidae

Thysanoptera - Thrips

Neuroptera - Alderflies, Dobsonflies,

    Fishflies, Snakeflies,

    Lacewings, Antlions, and

    Owlflies
    F. Chrysopidae
    F. Sialidae

Coleoptera - Beetles
    F. Carabidae
    F. Chrysomelidae
    F. Coccinellidae
    F. Dytiscidae
    F. Elateridae
    F. Gyrinidae
    F. Lampyridae
    F. Meloidae
    F. Ptinidae
    F. Scarabaeidae

Strepsiptera - Twisted-Wing Parasites

     F. Stylopidae

Mecoptera - Scorpionflies and

  Hangingflies

Siphonaptera - Fleas

     F. Ceratophyllidae

Diptera - Flies

     F. Asilidae
     F. Bombyliidae
     F. Calliphoridae
     F. Ceratopogonidae
     F. Culicidae
     F. Drosophilidae
     F. Empididae
     F. Heleomyzidae
     F. Oestridae
     F. Piophilidae
     F. Pipunculidae
     F. Sarcophagidae
     F. Simuliidae
     F. Syrphidae
     F. Tabanidae
     F. Tipulidae

Trichoptera - Caddisflies

Lepidoptera - Butterflies and Moths

     F. Arctiidae
     F. Choreutidae
     F. Danaidae
     F. Geometridae
     F. Heliconiidae
     F. Lasiocampidae
     F. Lycaenidae
     F. Lymantridae
     F. Noctuidae
     F. Nymphalidae
     F. Psychidae
     F. Saturnidae
     F. Sphingidae
     F. Tortricidae

Hymenoptera - Sawflies, Ants, Bees, and

  Wasps

     F. Apidae
     F. Colletidae
     F. Encyrtidae
     F. Eumenidae
     F. Formicidae
     F. Meloponidae
     F. Scelionidae
     F. Sphecidae
     F. Vespidae

REFERENCES FOR INSECTS DID IT FIRST

Akre, R. D. 1982. Social Wasps. Chapt. 1. pp. 1-105. In H. R. Herman, Ed., Social Insects. Vol. 4. Academic: NY 385 p.

Akre, R. D., A. Greene, J. F. MacDonald, P. J. Landolt, and H. G. Davis. 1981. The yellowjackets of America north of Mexico. USDA Handbook No. 552. 102 p.

Aiken, R. B. 1985. Sound production by aquatic insects. Biol Rev. 65:163-211.

Alcock, J. 1972. The evolution of the use of tools by feeding animals. Evolution 26: 464-473.

Anderson, S. and T. Weis-Fogh. 1964. Resilin. A rubberlike protein in arthropod cuticle. pp. 1-65. In J. Beament. J. Treherne, and V. Wigglesworth, Eds. Advances in Insect Physiology. Vol 2. Academic: NY. 364 p.

Atkins, M.D. 1980. Introduction to Insect Behavior. MacMillan: New York 237 p.

Barber, J. T., E. K. Ellgaard, L. B. Thien, and A. E. Stack. 1989. The use of tools for food transportation by the imported fire ant, *Solenopsis invicta*. Anim. Beh. 38(3):550-552.

Batra, S.W.T. and L.R. Batra. 1967. The fungus gardens of insects. Sci. Amer. 217(5): 112-120.

Bell, and R.T. Carde (Ed.) 1984. Chemical Ecology of Insects. Sinauer: Sunderland, Mass. 524 pp.

Bennet-Clark, H. C. and A. W. Ewing. 1970. The love song of the fruit fly. Sci. Amer. 223(1): 84-92.

Beroza, M. 1971. Insect sex attractants. Amer. Sci. 59: 320-325.

Birch, M. C., Ed. 1974. Pheromones. American Elsevier: NY. 495 p.

Blakemore, R. P. amd R. B. Frankel. 1981. Magnetic navigation in bacteria. Sci.Amer. 245:58-65.

Blum, M. S. 1981. Chemical Defenses of Arthropods. Academic: NY. 538 p.

Bonanos, C. 1992. The father of modern bridges. Am. Hert. Sci. Tech. 8: 8-20.

Borror, D.J., C.A. Triplehorn, and N. F. Johnson. 1989. An Introduction to the Study of Insects. Saunders: Philadelphia 875 p.

Borror, D.J. and R.E. White. 1970. A field guide to the insects of America North of Mexico. Houghton Mifflin: Boston. 404 p.

Bowers, W.S., L.R. Nault, R.E.Webb, and S.R. Dutky. 1972. Aphid alarm pheromone: isolation, identification, synthesis. Science 177: 1121-1122.

Brockman, H. J. 1985. Tool use in digger wasps (Hymenoptera: Sphecinae). Psyche 92: 309-329.

Brower, L. P. 1969. Ecological chemistry. Sci. Amer. 220: 22-29.

Bruce, J.A. and J.E. Carico. 1986. Silk use during mating in *Pisaurina mora* (Walckenaer) (Araneae, Pisauridae). J. Arach. 16(1): 1-4.

Buck, J. B. and E. M. Buck. 1976. Synchronous fireflies. Sci. Amer. 234: 74-85.

141

Carlson, A. D. and J. Copeland. 1978. Behavioral plasticity in the flash communication systems of fireflies. Am. Sci. 66: 340-346.

Carlson, A. D. and J. Copeland. 1985 Communications in insects. Quarterly. Rev. Biol. 60(4): 415-436.

Carthy, A. D. 1958. An introduction to the behaviour of invertebrates. Allen and Unwin, London. 380 p.

Catts, E. P. 1967. Biology of a California rodent bot fly *Cuterebra latifrons* Coq. J. Med. Entomol. 4: 87-101.

Chapman, R. F. 1982. The Insects: Structure and Function. 3rd Ed. Harvard University: Cambridge, Mass. 919 p.

Cole, F. R. 1969. The Flies of Western North Ameria. Univ. Calif. Press: Berkeley. 693 p.

Cott, H. B. 1966. Adaptive Coloration in Animals. Methuen: London. 508 p. +48 plates.

Cowen, R. 1990. Parasite power. Sci. News 138: 200-202.

Dixon, A. F. G. 1959. An experimental study of the searching behavior or the predatory coccinellida beetle *Adalia decaempunctata* (L.). J. Anim. Ecol. 28:259-281.

Dodge, H. R. 1966. Some new or little-known Neotropical Sacrophagidae (Diptera), with a review of the genus *Oxysarcodexia.* Ann. Ent. Soc. Amer. 59(4): 674-701.

Dunning, D. C. and K. D. Roeder. 1965. Moth sounds and the insect-catching behavior of bats. Science 147(3654):173-174.

Duplaix, N. 1988. Fleas: the lethal leapers. Nat. Geographic 173(5): 672-694.

Dyer, F. and J.L. Gould. 1983. Honey bee navigation. Amer. Sci. 71: 597.

Edwards, R. 1980. Social Wasps: Their biology and control. Rentokil: East Grinstead, England. 398 p.

Eisner, T., E. Van Tassell, and J. E. Carrel. 1967. Defensive use of a fecal shield by a beetle larva. Science 158: 1471-1473.

Erlich, P. R. and P. H. Raven. 1967. Butterflies and plants. Sci. Amer. 216: 104-113.

Elzinga, R. J. 1987. Fundamentals of Entomology. 3rd Ed. Prentice-Hall: Englewood Cliffs, NJ. 456 p.

Englemann, F. 1970. The Physiology of Insect Reproduction. Pergamon: NY. 307 p.

Esch, H. 1967. The evolution of bee language. Sci. Amer. 216:96-104.

Ewing, L. S. 1967. Fighting and death from stress in a cockroach. Science 155:1035-1036.

Fellers, J. H. and G. M. Fellers. 1976. Tool use in a social insect and its implications for competitive interactions. Science 192: 70-72.

Fowler, H. G. 1982. Tool use by *Aphaenogaster* ants: a reconsideration of its role in competitive interactions. Bull. New Jersey Acad. Sci. 27: 81-82.

Frost, S. W. 1959. Insect Life and Insect Natural History. 2nd revised ed. Dover: NY 526 p.

Gerber, H. S. and E. C. Klostermeyer 1970. Sex control by bees: a voluntary act of egg fertilization during oviposition. Science 167:82-84.

Gillott, C. 1980. Entomology. Plenim:NY 729 p.

Glancey, M., C. E. Stringer, Jr., C. H. Craig, P. M. Bishop, and B. B. Martin. 1973. Evidence of a replete caste in the fire ant, *Solenopsis invicta*. Ann. Ent. Soc. Amer. 66:233-234.

Goff, M. L., A. I. Omori, and J. R. Goodbrod. 1989. Effect of cocaine in tissues on the developmental rate of *Boettcherisca peregrinae* (Diptera: Sarchphagidae). J. Med. Ent. 26(2):91-93.

Gould, J. L. 1982. Ethology: The mechanisms and evolution of behavior. Norton: NY. 544 p. + 61 p. Appendix.

Gould, J. L. and C. G. Gould. 1988. The Honey Bee. Sci. Amer Lib.:NY. 239 p.

Gotwald, K. 1986. The beneficial economic role of ants. Chapt. 11, pp. 290-313. In S. B. Vinson, Ed., Economic Impact and Control of Social Insects. Praeger: NY. 421 p.

Hansen, L. D. and R. D. Akre. 1985. Biology of carpenter ants in Washington State (Hymenoptera:Formicidae:*Camponotus*). Melanderia 43:1-62.

Harwood, R. F. and M. T. James.1979. Entomology in Human and Animal Health. Macmillian:NY. 548 p.

Haskell, P.T. 1964. Sound production. pp 363-608. In: M. Rockstein (ed.) The Physiology of the Insecta. 2nd Ed. Vol. II Academic: New York. 640 pp.

Hefetz, A, H. M. Fales, and S. W. T. Batra. 1979. Natural polyesters: Dufour's gland macrocyclic lactones form brood cell laminesters in Colletes bees. Science:204:415-417.

Heinze, J. 1990. Dominance behavior among ant females. Naturwissenschaften 77:41-43.

Henderson, G. and R.D. Akre. 1986a. Biology of the myrmecophilous cricket, *Myrmecophila manni* (Orthoptera: Gryllidae). J. Kansas ent. Soc. 59(3):454-467.

Henderson, G. and R. D. Akre. 1986b. Dominance hierarchies in *Myrmecophila manni* (Orthoptera: Gryllidae). Pan-Pac. Ent. 62(1): 24-28.

Howard, R. W., R. D. Akre, and Wm B. Garnett. 1990. Chemical mimicry in an obligate predator of carpenter ants (Hymenoptera: Formicidae). Ann. Ent. Soc. Amer. 83:607-616

Hill, K. and A. M. Hurtado. 1989. Hunter-gatherers of the New World. Amer. Sci. 77(5): 436-443.

Holldobler, B.K. and E.O. Wilson. 1977. Weaver ants. Sci. Amer. 237(6): 146-154.

Holldobler, B. and E.O. Wilson. 1983. The evolution of communal nest-weaving in ants. Amer. Sci. 71:490-499.

Hölldobler, B. and E. O. Wilson. 1990. The Ants. Belknap/Harvard Univ. Press: Cambridge, MA. 732 p.

Huber, P. 1802. Observations on several species of the genus *Apis*, known by the name humble-bees, and called Bobinatrices by Linnaeus. Trans. Linn. Soc. Lond. Zool. 6: 214-298.

Hutchins, R. E. 1980. Nature Invented It First. Dodd, Mead: NY.111 p.

Ito, Y. 1960. Territorialism and residentiality in a dragonfly, *Orthetrum albistylum speciosum* Uhler (Odonata: Anisoptera). Ann. Entomol. Soc. Am. 53:851-853.

Jacobs, M. E. 1955. Studies on territorialism and sexual selection in dragonflies. Ecol. 36: 566-586.

Jacobson, M. 1971. Insect sex attractants. Amer. Sci. 59:320-325.

Jacobson, M. 1972. Insect Sex Pheromones. Academic: NY. 382 p.

Janvier, H. 1963. La mouche de la truffle (*Helomyza tuberiperda* Rondani). Bull. Soc. Ent. France 68(5/6):140-147.

Janzen, D. H. 1966. Coevolution of mutualism between ants and acacias in Central America. Evolution 20: 249-275.

Joklik, W. K. and H. P. Willett, Eds. 1976. Zinsser Microbiology. 16th Ed. Appleton-Century-Crofts: NY. 1223 p.

Kato, M. and L. Hayasaka. 1958. Notes on the dominance order in experimental populations of crickets. Ecol. Rev. 14:311-315.

Kerkut, G. A. and L. I Gilbert (Eds.). 1985. Comprehensive Insect Physiology, Biochemistry, and Pharmacology. Vol. 1-13. Pergamon: Oxford, England. 8126 p.

Kessel. E. L. 1955. The mating activities of balloon flies. Syst. Zool. 4:997-1004.

Kettlewell, H.B.D. 1965. Insect survival and selection for pattern. Science 148: 1290-1296.

Kettlewell, H.B.D. 1973. The Evolution of Melanism. Clarendon: NY. 424 p.

Kim, K. C. and R. W. Merritt (Eds.). 1987. B lackflies: Ecology, population management, and annoted world list. Penn. State Univ.: Univ. Park. 528 p.

Kirk, V. M. and B. J. Dupraz. 1972. Discharge by a female ground beetle, *Pterostichus lucublandus* (Coleopter: Carabidae), used as a defense against males. Ann. Ent. Soc. Am. 65: 513.

Kistncr, D. H. 1982. The Social Insects' Bestiary. Chapt. 1, pp.1-244. In H. R. Herman, Ed. Social Insects, Vol. 3. Academic: NY. 459p.

Laboulbene, J. A. 1864. Observations sur les insectes Tuberivores. Ann. Soc. Ent. France., Ser. 4 4:69-114.

Lall, A. B., H. H. Seliger, W. H. Biggley, and J. E. Lloyd. 1980. Ecology of colors of firefly bioluminescence. Science 210:560-562.

Lin, N. 1964-1965. Tho uso of sand grains by thc pavcment ant, *Tetramorium caespitum*, while attacking Halictine bees. Bull. Brooklyn Ent. Soc. 59-60: 30-34.

Liu, G. 1939, Some extracts from the history of entomology in China. Psyche 46:23-28.

Lipske, M. and K.B. Sandved. 1988. Letter perfect. Nat'l. Wildlife. 26(2): 12-13.

Lloyd, J. E. 1965. Aggressive mimicry in *Photuris*: firefly femmes fatales. Science 149:653-654.

Lloyd, J. E. 1971. Bioluminescent communication in insects. Ann. Rev. Ent. 16:97-122.

147

Lloyd, J. E. 1979. Mating behavior and natural selection. In: Symposium: Sociobiology of Sex. Florida Entomologist 62(1): 17-23.

Lofgren, C. S. , and R. K. Vander Meer, Eds.. 1986. Fire Ants and Leaf-Cutting Ants. Westview: Boulder, Colorado. 400 p.

Luscher, M. 1961. Air-conditioned termite nests. Sci. Amer. 205:138-145.

Mackerras, I. M. (Ed.). 1970. The Insects of Australia. Melbourne Univ. Press: Carlton, Victoria. 1029 p.

Maschwitz, U. and E. Maschwitz. 1974. Platzende arbeiterinnen: Eine neue art der Feindabwehr bei sozialen hautfluglern. Oecologia 14: 289-294.

Matheson, R. 1957. A Laboratory Guide in Entomology. Comstock: Ithaca, NY. 135 p.

Matthews, R. W. 1968. Nesting biology of the social wasp Microstigus comes (Hymenoptera: Specidae: Pemphredoninae). Psyche 75: 23-45.

McDonald, P. 1984. Tool use by the ant, *Novomessor albisetosus* (Mayr). J. N.Y. Ent. Soc. 92: 156-161.

McElroy, W. D. 1964. Insect Bioluminescence. Chapt. 11. pp.463-508. in M. Rockstein, Ed., The Physiology of the Insecta. Vol. 1. Academic: NY 640 p. .

McGovern, J. N., R. L. Jeanne, and M. J. Effland. 1988. The nature of wasp nest paper. Tappi Journal. December:133-139.

McKay, E. A. 1988. Tunneling to New York. Amer. Heritage of Invention and Technology. 4(2): 22-31.

McMahan, E. A. 1983. Bugs angle for termites. Nat. Hist. 92: 40-46.

Merritt, R.W. and J.B. Wallace. 1981. Filter-feeding insects. Sci. Amer. 244(4): 132-147.

Metcalf, C. L., W. P. Flint, and R. L. Metcalf. 1962. Destructive and Useful Insects. 4th Ed. McGraw-Hill: NY. 1087.

Michener, C. D. 1970. Hymenoptera, section on bees. p. 867-959. 1970. The Insects of Australia. Melbourne University Press, Carlson, Victoria, Australia 3053. Pp 1029.

Miller, L.A. and E.G. MacLeod. 1966. Ultrasonic sensitivity: a tympanal receptor in the green lace wing *Chrysopa carnea*. Science 154: 891-893.

Miller, N. C. E. 1971. The Biology of the Heteroptera. 2nd Ed. Classey: Hampton Middlesex: England. 206 p.

Milne, L. & M. Milne. 1980. The Audubon society field guide to North American insects and spiders. Alfred Knopf: NY. 989 p.

Moglich, M. H. J. and G. D. Alpert. 1979. Stone dropping by *Conomyrma bicolor* (Hymenoptera: Formicidae): a new technique of interference competition. Behav. Ecol. Sociobiol. 6: 105-113.

Naumann, I. D., Chief Editor. 1991. The Insects of Australia. Cornell Univ. Press: Ithaca, NY. Vol. I, pp. 1- 542, Vol 2, pp. 543-1137.

Oldroyd, H. 1964. The Natural History of Flies. Norton: NY. 324 p.

Papageorgis, C. 1975. Mimicry in Neotropical butterflies. Amer. Sci. 63: 522-532.

Pardi, L. 1948. Dominance order in *Polistes* wasps. Physiol. Zool. 21:1-13.

Pasteur, G. 1982. A classificatory review of mimicry systems. Ann. Rev. Ecol. Syst. 13:169-199.

Penman D. R. and W. W. Cone 1974. Role of web, tactile stimuli, and female sex pheromone in attraction of male twospotted spider mites to quiescent female deotonymphs. Ann. Ent. Soc. Amer. 67(2): 179-182.

Petroski, H. 1992. The Britannia tubular bridge. Am. Sci. 80: 220-224.

Philip, C. B. 1931. The Tabanidae (horseflies) of Minnesota, with special reference their biologies and taxomony. Minn. Agr. Exp. Sta. Tech. Bull. 80: 1-132.

Pringle, J. W. S. 1975. Insect Flight. Oxford Biology Reader. No. 52. Oxford Univ. Press: Oxford, England. 16 p.

Rettenmeyer, C. W. 1963. Behavioral studies of army ants. Univ. Kansas Sci. Bull. 44(9): 281-465.

Rettenmeyer, C.W. 1970. Insect mimicry. Ann. Rev. Ent. 15:43-74.

Richards, O. W. 1978. The Social Wasps of the Americas Excluding the Vespinae. British Museum (Nat. Hist.): London. 580 p.

Richards, O. W. and R. G. Davies. 1977. Imm's General Textbook of Entomology. 10th Ed. Vol 1: Structure, Physiology and Development. John Wiley and Sons: NY. 418 p.

Ritland, D. B. and L. P. Brower. 1991. The viceroy butterfly is not a batesian mimic. Nature 350: 497-498.

Roeder, K. D. 1965. Moths and ultrasound. Sci. Amer. 212(4):94-102.

Roeder, K. D. and A. E. Treat. 1961. The detection and evasion of bats by moths. Amer. Sci. 49(2);135-148.

Romoser, W.S. 1981. The Science of Entomology. Macmillan: New York. 575 p.

Ross, H. H., C.A. Ross, and J. R. P. Ross. 1982. A Textbook of Entomology. John Wiley & Sons: New York 666 p.

Roth, L.M. and T. Eisner. 1962. Chemical defenses of arthropods. Ann. Rev. Ent. 7: 107-136.

Rothschild, Y. Schlein, K. Parker, C. Neville, and S. Sternberg. 1973. The flying leap of the flea. Sci. Amer. 229(5): 92-100.

Royer, M. 1975. Hermaphrodism in Insects. Studies on *Icerya purchasi.* Pp. 135-145. In R. Reinboth (Ed.) Intersexuality in the Animal Kingdom. Springer-Verlag: NY. 449p.

Sargent, T. D. 1966. Background selections of geometrid and noctuid moths. Science 154:1674-1675.

Saunders, D. S. 1976. The biological clock of insects. Sci. Amer. 234:114-121.

Saunders, D. S. 1982. Insect Clocks. 2nd Ed. Pergamon: New York. 409 p.

Schildknecht, H. 1970. The defensive chemistry of land and water beetles. Angew. Chem. 9(1):1-9.

Schildknecht, H. 1971. Evolutionary peaks in the defensive chemistry of insects. Endeavour 30: 136-141.

Schneirla, T. C. (Ed., H. R. Topoff). 1971. Army Ants: A study in social organization. Freeman: San Francisco. 349 p.

Schultz, G. W. 1982. Soil-dropping behavior of the pavement ant, *Tetramorium caespitum* (L.) (Hymenoptera: Formicidae) against the alkali bee. J. Kansas Ent. Soc. 55: 277-282.

Seyle, H. 1973. The evolution of the stress concept. Amer. Sci.61:692-706.

Sherman, H. 1989. Polyester repair-man is really a bee. Agric. Res. 37:18.

Shorey, H.H. 1973. Behavioral responses to insect pheromones. Ann Rev. Ent. 18: 349-380.

Shorey, H.H. 1976. Animal Communication by Pheromones. Academic: NY. 167 p.

Smith, K.G.V. 1986. A Manual of Forensic Entomology. Comstock: Ithaca, New York. 205 p.

Snodgrass, R. E. 1923. Insect musicians, their music, and their instruments. Smithsonian Annual Report for 1923. 1923:405-452

Spradbery, J. P. 1973. Wasps: An account of the biology and natural history of solitary and social wasps. Univ. Washington Press: Seattle. 408 p.

Steiner, A. L. 1983. Predatory behavior of digger wasps (Hymenoptera: Sphecidae) VI. Cutworm hunting and stinging by the ammophiline wasp *Podalonia luctuosa* (Smith). Melanderia 41:1-16.

Stolzenburg, W. 1990. When life got hard. Science News 138:121-122,123.

Strassman, J. E. 1979. Honey caches help female paper wasps (*Polistes annularis*) survive Texas winters. Science 204: 207-209.

Suomi, D. 1988. Snailcase bagworm. Cooperative Extension, Wash. State Univ. Ext. Bull. 1485 2 p.

Talou, T., A. Gaset, M. Delmas, M. Kulofaj, and C. Montant. 1990. Diethyl sulphide: the secret for black truffle hunting by animals? Mycol. Res. 94: 277-278.

Thornhill, R. and J. Alcock. 1983. The Evolution of Insect mating Systems, Harvard: Cambridge, Mass. 547 p.

Topoff, H. 1984. Invasion of the body snatchers. Nat. Hist. 93(10):78-84.

Torchio, P. F., and D. J. Burdick. 1988. Comparative notes on the biology and development of *Epeolus compactus* Cresson, a cleptoparasite of *Colletes kincaidii* Cockerell (Hymenoptera, Colletidae). Annals Entomol. Soc. Amer. 81: 626-636.

Torchio, P. F., G. E. Trostle, and D. J. Burdick. 1988. The nesting biology of *Colletes kincaidii* Cockerell (Hymenoptera, Colletidae) and development of its immature forms. Annals Entomol. Soc. Amer. 81: 605-625.

Tucker, V. A. 1969. Wave making by whirligig beetles (Gyrinidae). Science 166:897-899.

Van der Pijl, L. and C. H. Dodson. 1966. Orchid Flowers. Fairchild Topical Garden/Univ. Miami Press, Coral Gables. 214 p.

Von Frisch, K. 1966. The Dancing Bees: An Account of the Life and Senses of the Honey Bee. Methuen: London. 198 p.

Von Frisch, K. 1967. The Dance Language and Orientation of Bees. Belknap/Harvard Univ. Press:Cambridge, Mass. 566 p.

Von Frisch, K 1983. Animal Architecture. Van N- 'rant Reinhold: NY. 306 p.

Walker, T. J. and D. Dew. 1972. Wing movements of calling katydids: fiddling finesse. Science 178:174-176.

Weber, N.A. 1972. Gardening Ants, the Attines. American Philosophical Society: Philadelphia, Pennsyl. 146 p.

Wehner, R. 1976. Polarized light navigation by insects. Sci. Amer. 235:106-115.

Weis-Fogh, T. 1975. Unusual mechanisms for the generation of lift in flying animals. Sci. Amer. 233(5): 80-87

Wenner, A. M. 1964. Sound communication in honeybees. Sci. Amer. 210(4): 117-123.

West, M. J. 1967. Foundress association in Polistine wasps: Dominance hierarchies and the evolution of social behavior. Science 157: 1584-85.

Wheeler, W.M. 1910. Ants Their Structure, Development and Behavior. Columbia University: New York. 663 p.

Wheeler, W.M. 1973. The Fungus Growing Ants of North America. Dover: New York. 136 p.

Wigglesworth, V. B. 1972. The Principles of Insect Physiology. 7th Ed. Chapman and Hall: London. 827 p.

Wilson, E. O. 1963. Pheromones. Sci. Amer. 208:100-114.

Wilson, E.O. 1971. The Insect Societies. Belknap/Harvard Univ. Press: Cambridge. 548 p.

Wilson E.O. 1975. Slavery in ants. Sci. Amer. 232(6): 32-36.

Winston, M. L. 1987. The Biology of the Honey Bee. Harvard Univ. Press:Cambridge, Mass. 281 p.

Wynne-Edwards, V. C. 1962. Animal Dispersion in Relation to Social Behavior. Oliver and Boyd: London. 653 p.

INDEX TO CHAPTERS

158

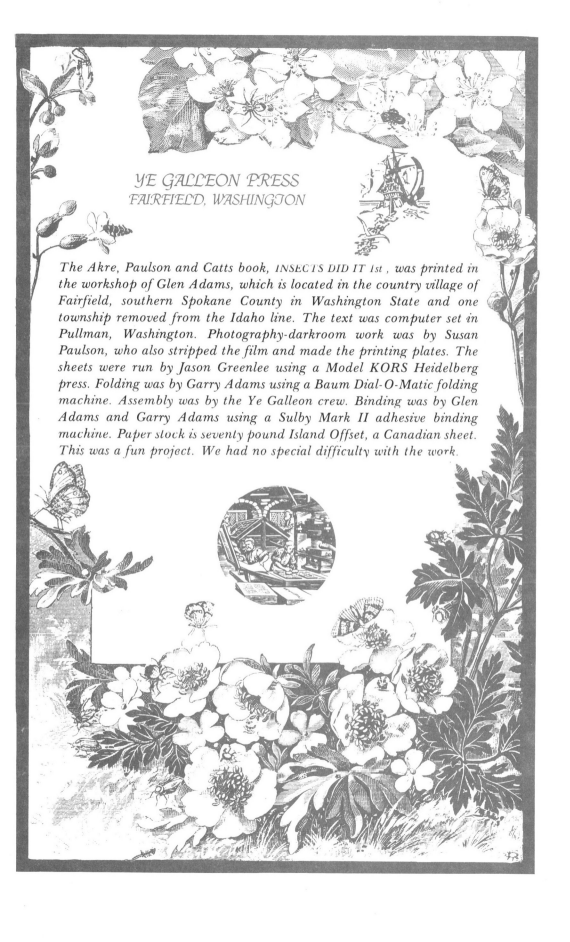

YE GALLEON PRESS
FAIRFIELD, WASHINGTON

The Akre, Paulson and Catts book, INSECTS DID IT 1st , was printed in the workshop of Glen Adams, which is located in the country village of Fairfield, southern Spokane County in Washington State and one township removed from the Idaho line. The text was computer set in Pullman, Washington. Photography-darkroom work was by Susan Paulson, who also stripped the film and made the printing plates. The sheets were run by Jason Greenlee using a Model KORS Heidelberg press. Folding was by Garry Adams using a Baum Dial-O-Matic folding machine. Assembly was by the Ye Galleon crew. Binding was by Glen Adams and Garry Adams using a Sulby Mark II adhesive binding machine. Paper stock is seventy pound Island Offset, a Canadian sheet. This was a fun project. We had no special difficulty with the work.